THE INTELLIGENCE ADVANTAGE

How Business Leaders Win in the Age of AI

First Edition · 2026

Samuel S. Conn, Ph.D.

The Intelligence Advantage: How Business Leaders Win in the Age of AI

 Printed in the United States of America

Published by Business Academic Press

ISBN: 979-8-9964689-2-8

First Edition, 2026

The GenAI Divide…

MIT NANDA (2025) studied hundreds of enterprise AI deployments and found that roughly ninety-five percent failed to produce any measurable financial return, despite tens of billions of dollars in enterprise investment. The difference between the rare successes, and an overwhelming majority of failures, had little to do with model quality. It came down to whether the AI system was built to keep learning (to retain feedback, adapt to context, and improve on its own) or whether it was simply a static tool bolted onto an unchanged workflow. The organizations in the successful five percent had built exactly what this book argues is the real source of advantage: systems that compound.

MIT NANDA. (2025). *The GenAI divide: State of AI in business 2025.* MIT Media Lab.

TABLE OF CONTENTS

FOREWORD

I HAVE SPENT THE BETTER part of two decades watching executives make technology bets, and I have learned to be suspicious of books that promise a shortcut through complexity. This is not that book.

What Samuel Conn has written instead is something rarer: a guide that respects how much genuine intellectual work is required to turn artificial intelligence from a line item into a source of durable advantage. He does not ask you to take AI on faith, and he does not ask you to become an engineer. He asks something harder... that you become literate enough in the technology to govern it, strategic enough to prioritize it correctly, and patient enough to survive the trough before the payoff arrives.

That patience is, I think, the book's quiet argument. Long before most executives had heard the term, this book had already internalized what Erik Brynjolfsson's research on general-purpose technologies predicts: that the value of a transformation like this one does not arrive on the schedule of the technology. It arrives on the schedule of the organization, namely its workflows, its data foundations, its willingness to redesign itself rather than simply automate what it already does. Every enterprise now living through the gap between AI's promise and its measured return will recognize this book's diagnosis of that gap, and will find, chapter by chapter, the actual work required to close it.

That work is unglamorous, which is exactly why so few books cover it properly. Data architecture. Cloud economics. The build-buy-partner decision. The question of who, precisely, owns an AI system's failures when it has none of the accountability structures a human employee would have. Dr. Conn does not skip past these questions to get to the exciting part. He treats them as the exciting part, because he understands, correctly, that they are where AI initiatives succeed or quietly die.

I found myself most persuaded by the sections other authors tend to avoid entirely: the honest accounting of why AI projects fail even after the pilot succeeds, the case for treating data and model advantages as genuine competitive moats rather than marketing language, and the argument that governance is not a brake on innovation but the structural capability that lets innovation compound safely over time. Few business books are willing to sit with failure long enough to learn its actual mechanics. This one does.

If you are a leader who has been handed responsibility for your organization's AI strategy (whether that responsibility came with a title like Chief AI Officer or simply arrived on your desk unannounced) you will find in these pages what you actually need: not certainty, which no one can honestly offer you right now, but a map of the terrain and a disciplined way of thinking about it. That is a rarer gift than it sounds, and it is the reason I recommend this book without reservation to anyone standing where I once stood, uncertain how to lead through a transformation that had already begun.

Christopher Markham, Ph.D.

PREFACE

WE STAND AT A UNIQUE juncture in the history of business. The forces that drove past industrial revolutions, from steam power to the internet, are being superseded by a new and more profound engine of change: Artificial Intelligence. AI is not merely a tool for optimization; it is a fundamental shift in how value is created, decisions are made, and competition is waged. Ignoring this transition is no longer a viable strategic option. The Intelligence Advantage: How Business Leaders Win in the Age of AI is your essential guide to navigating this new reality, written with a single purpose: to equip you with the literacy, strategy, and practical knowledge required to integrate AI into your organization... not as a passing technology, but as a core competency.

This book was written for a diverse audience united by a common goal: harnessing AI for tangible business value. Current and aspiring business leaders will find the strategic literacy needed to invest in, champion, and govern AI initiatives. Business analysts and consultants will gain the structured knowledge to identify, scope, and execute profitable AI projects. Students of business, technology, information systems, and data science will find a bridge between academic theory and the realities of deploying AI in a corporate environment. A common pitfall in enterprise AI adoption is the chasm between the boardroom and the engineering team: leaders understand the need for AI but often lack the vocabulary to scope projects effectively, while technical teams may build impressive models that fail to solve high-value business problems. This textbook is designed to close that gap, giving managers and strategists the fluency to identify opportunities with real ROI and giving technologists and analysts the business context to align their work with organizational goals.

The use of AI platforms is changing rapidly as multi-model AI orchestration platforms emerge. Not entirely proven effective or given coverage in this book, these emerging platforms serve to improve the efficient and effective use of all available AI models. Multi-model AI orchestration is the practice of routing a single task across several different AI models rather than relying on one model to do everything, for example: a text-reasoning model might handle planning and copywriting, an image model handles visuals, a video model handles motion, and a separate coordinating layer decides which model gets which piece of the job and stitches the results back together. It's achieved through an *orchestration layer* (sometimes called a router, an agent, or a "brain") that sits above the individual models, breaks an incoming request into sub-tasks, matches each sub-task to whichever model is best suited for it, passes context and constraints along consistently, and then assembles the outputs into a single finished result. The appeal is straightforward: no *single model* is best at everything, and new ones supersede old ones every few months, so building a system that can swap in whichever model currently performs best (without the user having to know or care which one that is) turns a fragmented, fast-moving model landscape into something that behaves like one coherent tool.

The journey through this book is structured into four progressive parts, mirroring the maturation of an AI-ready organization: Part 1 establishes a shared vocabulary and builds the business case for investment; Part 2 demystifies the core technologies, from supervised machine learning to computer vision; Part 3 details the practical steps of the AI Project Life Cycle, from data preparation to deployment and scaling; and Part 4 tackles the critical issues of ethics, fairness, and governance that keep your strategy sustainable as the regulatory landscape evolves. AI is not a trend… it is the new infrastructure of commerce. By mastering the concepts in these pages, you will be prepared not just to participate in the future of business, but to lead it.

INTRODUCTION

Why This Book, Why Now, and Why It Is Structured the Way It Is

WE ARE LIVING THROUGH A moment of technological transformation that arrives, by most serious assessments, once in a generation… the kind of moment that separates, in retrospect, the organizations that recognized what was happening and responded with strategic seriousness from those that underestimated, delayed, or misunderstood the nature of the change until the cost of catching up exceeded the resources available to do so. Artificial Intelligence (AI) is not a feature upgrade to existing business systems, a productivity tool to be evaluated on the same terms as prior enterprise software investments, or a competitive amenity that organizations can adopt at their leisure once the technology matures and the risks become clearer. AI exists as a general purpose technology of the kind that economists identify as genuinely transformative… a technology that, like electricity, the internal combustion engine, and the internet before it, does not simply improve existing ways of doing things but fundamentally changes what is possible, who can do it, at what cost, and at what speed. The enterprises that internalize this distinction early (that treat AI not as a technology decision to be managed by their IT function but as a strategic transformation to be led by their most senior decision-makers) will define the competitive landscape of the next decade. Those that do not will find themselves navigating that landscape on terms set by others. This book is written for leaders who intend to be among the former, and it is written with the conviction that the difference between those two outcomes is not primarily a function of access to technology, capital, or talent (all of which are more widely available than the current discourse suggests) but a function of understanding: understanding what AI actually is and is not, what it can and cannot do, what the organizational and economic conditions for its successful deployment look like, and what the strategic, governance, and human leadership requirements of genuine AI transformation demand of the people responsible for guiding their organizations through it.

The audience this book is written for is specific and the specificity is intentional. Business executives and technology leaders who carry responsibility for AI-related decisions (investment decisions, strategic decisions, organizational decisions, governance decisions) without necessarily having the technical background to evaluate those decisions from first principles need a resource that meets them where they are: sophisticated in their understanding of business strategy, organizational dynamics, and economic value creation, but potentially without the deep technical fluency in machine learning, data architecture, and AI system design that would allow them to engage critically with the technical dimensions of the decisions they must make. Graduate students in business and information systems programs who are preparing to enter or advance within organizations undergoing AI transformation need a resource that bridges the gap between the technical AI education they receive in quantitative courses and the strategic, organizational, and governance dimensions of AI deployment that will define the nature and quality of their professional contribution. Technology executives (i.e., chief information officers, chief data officers, chief technology officers, and the senior technology leaders who report to them) need a resource that connects the technical architecture decisions they own to the business strategy, competitive dynamics, economic frameworks, and governance requirements that give those decisions their organizational context and their strategic significance. This book is written to serve all three audiences simultaneously, and the way it achieves that ambition is through a consistent commitment to analytical depth without unnecessary technical complexity… treating its readers as the intellectually capable, organizationally experienced, and strategically motivated professionals they are, and engaging the full complexity of AI transformation without retreating into either the technical jargon that excludes non-specialists or the executive summary superficiality that fails to provide the depth of understanding that consequential decision-making requires.

The architecture of this book is not accidental but rather designed around a specific and carefully considered organizing principle: the ***arc of executive decision-making*** that AI transformation requires, from initial understanding through strategic development, economic justification, competitive positioning, technical and organizational building, deployment and scaling, functional transformation, governance and compliance, leadership, and forward-looking strategic anticipation. This *arc*, defined by the logical progression to *understand, strategize, justify, compete, build, deploy, transform, govern, lead, and look ahead* maps directly onto the chapter sequence of the book, and understanding this mapping is the key to understanding why the book is organized as it is and how to use it most effectively. The book begins, in Chapters 1 and 2, with the understanding phase: establishing a working definition of AI that is precise enough to be analytically useful without being technically intimidating, tracing the historical development that produced the current AI landscape, and providing the conceptual framework for the core AI technologies (machine learning, deep learning, natural language processing, computer vision, and the emerging category of agentic AI) that business leaders must understand well enough to evaluate, govern, and strategically deploy without necessarily being able to build. Chapter 3 then performs a function that most AI business books conspicuously avoid: an honest, rigorous, and detailed examination of why AI initiatives fail (technically, organizationally, economically, and humanly) positioned early in the book because understanding failure modes before examining success frameworks is the analytical sequence that produces the most durable and practically useful understanding. A leader who knows what success looks like without understanding what failure looks like is equipped to recognize the destination but not to navigate the hazards that lie between the starting point and the arrival.

The strategize and justify phases of the executive decision arc are addressed in Chapters 4 and 5, which together provide the strategic and economic framework that AI investment decisions require. Chapter 4 develops the AI strategy framework… use case identification and prioritization, build versus buy versus partner decision architecture, AI portfolio construction, maturity modeling, and the organizational design questions that AI strategy inevitably raises. Chapter 5 addresses the economics of AI with the rigor that financial decision-makers require: the new cost structures of AI-era technology investment, the valuation frameworks appropriate for AI projects and AI-intensive businesses, the role of intangible capital in AI ROI, and the financial governance tools that allow organizations to manage AI investment portfolios with the same discipline they bring to capital allocation decisions in other domains. Chapters 6 through 9 address the compete, build, deploy, and transform phases: the competitive strategy analysis of AI moats and industry disruption patterns in Chapter 6, the data and infrastructure foundations that AI deployment requires in Chapter 7, the scaling, organizational design, and change management challenges of enterprise AI deployment in Chapter 8, and the functional transformation that AI enables across marketing, finance, operations, human resources, and innovation in Chapter 9. These four chapters constitute the operational core of the book (the section most directly relevant to the technology executives and program leaders responsible for the practical execution of AI transformation) and they are written with sufficient technical and organizational depth to serve as genuine working references for the decisions and challenges that execution entails.

The govern and lead phases of the executive arc are addressed in Chapters 10 and 11, which together provide the governance, compliance, and leadership framework that sustainable AI transformation requires. Chapter

10 examines responsible AI, governance, and regulation with the comprehensiveness that the current regulatory environment demands: the EU AI Act and its global regulatory parallels, intellectual property and data sovereignty, explainability and bias mitigation, model risk management, and the enterprise AI governance frameworks that allow organizations to move ambitiously in AI while managing the legal, ethical, and reputational risks that ungoverned AI deployment generates. The placement of governance after the functional transformation chapter is deliberate… governance is most meaningfully understood in the context of the specific AI deployments it is designed to govern, and the reader who encounters Chapter 10 having worked through the preceding chapters will engage its frameworks with the operational and strategic context that makes them analytically meaningful rather than abstractly procedural. Chapter 11 addresses the leadership dimension of AI transformation… building and retaining AI talent, managing organizational change, structuring teams for AI effectiveness, measuring AI program success against meaningful indicators of business value rather than technical performance proxies, and cultivating the organizational culture that makes AI transformation self-sustaining rather than dependent on the continuous intervention of senior leadership. Leadership is positioned second-to-last not because it is less important than the preceding chapters (it is arguably the most important) but because it is most effectively addressed in the context of the full strategic, technical, organizational, and governance complexity that the preceding chapters have established.

The book concludes in Chapter 12 with a look-ahead at the future of AI for business… an examination of the forces, technologies, and organizational dynamics that will shape enterprise AI over the next five years and beyond. This chapter addresses the Gartner AI Hype Cycle and its implications for enterprise investment timing, the trajectories of agentic AI and artificial general intelligence and their business implications, the emerging HTAP architectural paradigm that will define the next generation of AI-ready data infrastructure, the Productivity J-Curve that Erik Brynjolfsson's research predicts and that enterprises navigating AI transformation will inevitably encounter, and the characteristics of the AI-ready organization that will be best positioned to capture the compounding returns that AI transformation ultimately delivers for those who complete it fully and govern it wisely. Throughout all twelve chapters, the book maintains a consistent analytical posture that reflects its core conviction: that the most important questions in enterprise AI are not technical questions about which algorithms to use or which platforms to deploy, but strategic, organizational, economic, and human questions about how to build the institutional capability, governance infrastructure, and leadership culture that allow AI technology (whatever specific form it takes as the field continues to evolve) to generate genuine and durable business value. Technology changes faster than organizations can track. The frameworks for thinking about technology strategically, governing it responsibly, leading its adoption humanly, and building organizational capability that compounds over time are more stable, more transferable, and ultimately more valuable than any specific technical knowledge that the current state of AI can provide. That is what this book offers, and it is offered with the conviction that the leaders who engage with it seriously will be better equipped (not just better informed) for the transformation that is already underway and that will define the competitive landscape of the decade ahead.

CHAPTER 1

Defining AI and the Business Value Proposition

A Working Definition of AI for Business

ARTIFICIAL INTELLIGENCE (AI) has long been portrayed in popular culture as a futuristic force capable of surpassing human capability in every dimension - an idea that has shaped public perception for decades. Science fiction films, novels, and media often depict AI as sentient robots, omniscient supercomputers, or autonomous entities that rival or threaten humanity. While these portrayals capture the imagination, they also contribute to misconceptions about what AI currently is and what it can realistically achieve. In the business world, where many decisions must be grounded in feasibility and return on investment, such exaggerated expectations can create confusion. Thus, a practical definition of AI for business is essential - one rooted in real-world capabilities rather than dramatic speculation.

When we talk about AI in a business context, we are referring to technology that ***augments*** human abilities rather than replaces them wholesale. AI for business can be defined as a system of computer programs and algorithms designed to perform tasks that typically require human intelligence - tasks such as learning from data, recognizing patterns and behaviors, reasoning through information, perceiving environments, solving problems, and making decisions. What largely distinguishes business-focused AI from the broader, more philosophical interpretations is its purpose. In business organizations, AI is not pursued for novelty or theoretical exploration; it is implemented with a clear, measurable objective: to generate tangible business value.

This emphasis on measurable value is what makes the definition particularly crucial. In corporate environments, every technology investment must justify its existence through quantifiable outcomes. An AI system that does not reduce operational costs, increase efficiency, enhance customer experience, or create new revenue opportunities is not considered successful, no matter how advanced the underlying algorithms may be. The focus shifts from capability to contribution - how AI can support organizational goals, streamline processes, and empower teams to make better decisions?

Moreover, defining AI in this grounded way helps organizations avoid the trap of adopting technology for the sake of appearing innovative. This definition reframes AI not as a magical solution but as a strategic tool that depends on high-quality data, thoughtful implementation, and alignment with business strategy. Companies that treat AI as a silver bullet often underestimate the importance of domain expertise, change management, and continuous optimization. By contrast, those who view AI through the lens of measurable value tend to create systems that are more sustainable, scalable, and integrated with real business needs.

Another benefit of this practical definition is that it democratizes AI across organizations. When AI is presented as an accessible set of tools rather than an unreachable technological marvel, non-technical leaders can engage more confidently in discussions about its use. Marketing teams might explore AI-driven customer segmentation, operations teams might investigate predictive maintenance, and HR departments might consider AI-assisted talent analytics. This clarity fosters cross-functional collaboration, allowing AI to become a *shared resource* rather than a mysterious asset controlled exclusively by technical experts.

Ultimately, defining AI for business in clear, outcome-oriented terms helps organizations approach the technology with both enthusiasm and realism. Leaders are encouraged to ask better questions: What problem are we solving? How will we measure success? What data do we need? Who will use the system, and how will it fit into existing workflows and infrastructure? By grounding AI in present-day capability and business value, companies position themselves not only to adopt the technology effectively but also to innovate responsibly and sustainably. In doing so, they transform AI from an abstract concept into a practical driver of competitive advantage.

A Short History of AI (The Road to Practicality)

The concept of intelligent machines has fascinated thinkers for centuries, appearing in mythology, early mechanical inventions, and philosophical writings long before computers existed. Yet the formal discipline of Artificial Intelligence began in 1956 at the Dartmouth Summer Research Project, where a group of pioneering researchers proposed that aspects of human intelligence could be simulated by machines. This gathering marked the birth of AI as a scientific field and set the stage for decades of experimentation, optimism, and setbacks. To understand why AI has only recently become a central topic in business strategy, it is helpful to look at its evolution through three distinct historical waves.

The first wave, often called **the *Symbolic Era***, spanned the 1950s through the 1970s. During this period, researchers focused on symbolic reasoning… programming computers with explicit rules, logical structures, and human knowledge. Expert systems, one of the most notable achievements of this era, attempted to replicate the decision-making of domain specialists by encoding large libraries of "if–then" rules. While these systems showed promise in tightly controlled environments, they struggled with real-world unpredictability. They could not adapt, learn from experience, or handle exceptions well. As soon as the complexity of a situation exceeded the programmer's predefined rules, the system collapsed. While groundbreaking at the time, these early AI systems revealed the limitations of trying to hard-code intelligence directly into software.

The second wave, **the *AI Winters***, unfolded from the 1980s through the early 2000s. This era was characterized by disillusionment, reduced funding, and a dramatic slowdown in progress. Several factors contributed to these winters: computing power was insufficient for large-scale AI experimentation, access to (big) data was extremely limited compared to today's standards, and early AI systems had failed to deliver on the futuristic promises made by researchers and the media. Investors and institutions grew skeptical, leading to sharp declines in research budgets. The term "AI Winter" reflects the isolation and stagnation that settled over the field as excitement cooled and many projects were abandoned. Yet this period also served as a necessary reset, pushing researchers to reconsider assumptions and explore new directions.

The third wave, **the *Modern Era***, began in the 2010s and continues today. This period is marked by enormous breakthroughs, especially in machine learning, large language models and deep learning. Unlike symbolic AI, modern machine learning allows algorithms to detect patterns and improve performance through exposure to massive amounts of data. This *shift from rule-based programming to data-driven learning* has revolutionized the field. One of the primary reasons AI has surged into mainstream business relevance is the availability of *big data*… the explosion of digital records, social media content, click streams, sensor output, e-commerce transactions, and more. These vast datasets provide the raw material necessary to train today's high-capacity models.

Alongside big data, ***cloud computing*** has become a foundational driver of modern AI success. In the past, training complex models required expensive hardware that only specialized institutions could afford. Today, organizations of all sizes can access scalable computing resources, including high-performance Graphics Processing Units, through cloud platform providers. This accessibility has dramatically *lowered the barrier to entry*, allowing businesses to experiment with and deploy AI at unprecedented speed and scale. The "cloud" has transformed AI from a niche research activity into a practical tool for everyday business operations.

Finally, ***algorithmic advancements***, particularly in deep learning, have delivered performance leaps across a wide range of tasks. Refinements in neural network architectures (such as *convolutional networks* for image analysis and *transformer models* for language processing) have made it possible to achieve accuracy levels once thought unattainable. These breakthroughs underpin many modern applications: virtual assistants, recommendation systems, fraud detection tools, autonomous vehicles, and more. The achievements of this era are less about mimicking human consciousness and more about engineering systems that excel at specific tasks through computation, pattern recognition, and vast amounts of data.

Taken together, this historical progression shows that AI's current success is not the product of a sudden theoretical revelation but rather the result of steady advancements in engineering, hardware, and data availability. AI has emerged as a powerful business tool because the necessary conditions and resources - abundant data, affordable computing, and sophisticated algorithms - have finally converged. This perspective helps demystify AI's rise and underscores that today's capabilities are built on decades of incremental progress, experimentation, and refinement.

Categorizing AI: Narrow vs. General AI

To set realistic expectations around what AI can and cannot do, business organizations must distinguish between the technology that exists today and the speculative forms often portrayed in science fiction. Popular media tends to blur these lines, showing AI as fully autonomous, emotionally intelligent, and capable of performing an unlimited range of tasks. Today, the AI systems deployed across industries are far more specialized. Understanding this difference is not just an academic exercise but has direct implications for budgeting, planning, and setting achievable business goals within an organization.

Most of the AI in use today falls into the category of ***Narrow (or Weak) AI***, which refers to systems designed to perform one specific task, or a tightly defined set of tasks, with exceptional efficiency. These systems excel not because they "understand" in a human sense but because they process vast amounts of data far more quickly than humans can. Examples are everywhere: spam filters that identify unwanted emails, recommendation engines that suggest products or content, fraud detection algorithms that scan financial transactions for anomalies, and even the navigation systems supporting autonomous vehicles. Each of these solutions is highly capable within its intended domain. However, they *cannot pivot* to an unrelated function. A system trained to detect fraudulent credit card activity cannot compose music, diagnose illnesses, or interpret legal contracts. This task-specific limitation is a defining characteristic of narrow AI and is precisely why current AI is so reliable in controlled business environments and implementations.

Because of their specialization, narrow AI systems are extraordinarily powerful tools for solving well-defined business problems. They optimize operations, speed up decision-making, improve customer experiences, and unlock efficiencies that would be impossible through manual processes alone. Their precision is an advantage, not a weakness. When a company clearly defines a problem it wants to solve (whether it is reducing churn, optimizing supply chains, or automating routine workflows) narrow AI can deliver measurable, repeatable value. This alignment between the system's capabilities and the business organization's needs is what makes narrow AI transformative in practice.

In contrast, ***General (or Strong) AI*** refers to a theoretical class of systems capable of understanding, learning, and applying intelligence across any domain… essentially matching or exceeding human cognitive abilities. Such a system would not require task-specific training; this class of system would be capable of improvisation, reasoning, and flexible problem-solving across diverse contexts. While captivating in theory and frequently depicted in movies and literature, general AI arguably does not currently exist. Most agree that there is no system today (commercial, academic, or experimental) that can think or act with the broad adaptiveness of a human mind. As a result, general AI plays a limited role today in current business strategy, investment planning, or operational decision-making.

Clarifying the distinction between narrow and general AI helps managers avoid misinterpretations and inflated expectations. Without this understanding, leaders may assume that AI can autonomously handle ambiguous or creative tasks, or that it will evolve spontaneously into more advanced forms. Such misunderstandings can lead to poorly scoped projects, wasted resources, and disappointment when systems do not perform beyond their designed capabilities. By contrast, business organizations that anchor their strategy around the real, proven capabilities of narrow AI are better positioned today to build initiatives that succeed.

Ultimately, recognizing that modern AI is a powerful (but highly specialized) tool enables businesses to focus on achievable outcomes. Rather than waiting for hypothetical breakthroughs or chasing science-fiction visions, business organizations can invest in practical applications that drive measurable results today. This mindset helps business leaders set realistic project requirements, choose the right technologies, and build AI solutions that deliver tangible value across their operations.

The AI Value Chain: Data, Algorithms, and Computation

An effective AI system is not a single piece of "software" but an interconnected ecosystem made up of three essential components: data, algorithms, and computation. These elements operate together much like the legs of a three-legged stool… remove or weaken one and the entire structure becomes unstable. For organizations investing in AI, understanding this value chain is crucial. This knowledge enables business leaders to allocate resources wisely, anticipate risks, and design systems that perform reliably in real-world environments. When businesses overlook any one of these components, even the most promising AI initiatives can fail to deliver meaningful results.

The first and most foundational component is *data*, the raw material that fuels modern AI. Unlike traditional software, which follows predefined rules, AI systems learn by identifying patterns within large datasets. This makes the quality, quantity, and diversity of data central to the system's performance. Organizations must pay attention to several big data dimensions, namely: ***VOLUME*** - ensures that the model sees enough examples to generalize accurately; ***VELOCITY*** - determines whether incoming data can be processed quickly enough to support real-time or near-real-time decision-making; ***VARIETY*** - reflects the different types of data (e.g., text, numbers, images, sensor signals) that enrich the system's understanding; and finally, ***VERACITY*** - ensures the data is clean, accurate, and free from bias or noise. Weakness in any of these dimensions compromises the model's output, making investments in data governance and data engineering just as important as investments in the AI model and infrastructure.

The second component is the *algorithm*, often referred to as "the model." The model is the mathematical engine that transforms raw data into insights, predictions, or decisions. Algorithms range from simple statistical techniques like linear regression to highly sophisticated deep neural networks with millions (or even billions) of parameters. The choice of algorithm *must always begin* with the business problem: predicting customer churn, detecting fraudulent behavior, routing support tickets, or personalizing recommendations. Different problems demand different modeling approaches, and more complex is not always better. A simple model may outperform a complex model if it is better aligned with the structure and quality of the available data. The key is thoughtful model selection grounded in the practical needs of the organization.

The third component, *computation*, provides the hardware and technical infrastructure required to train, deploy, and maintain AI models. Modern AI, especially deep learning, demands immense processing power during the training phase. This typically involves Graphics Processing Units (GPUs) or Tensor Processing Units (TPUs), which can perform the parallel calculations necessary to optimize large models. As noted previously, cloud platforms have revolutionized access to this power, enabling organizations to experiment with sophisticated AI without building their own data centers. Once the model is trained, it must be deployed in an environment that can deliver predictions reliably and at the speed required by the business use case. For example, a customer service chatbot must generate responses instantly, and a fraud detection algorithm must evaluate transactions in milliseconds. Robust infrastructure ensures the model can operate consistently under varying loads and performance requirements.

Understanding these three components as an *integrated value chain* helps organizations visualize where AI initiatives might encounter obstacles. A business organization may have cutting-edge algorithms but insufficient data; another may possess rich datasets but lack the computational power to train advanced models; still another may excel at model development but struggle to deploy systems that scale. Each weakness introduces risk, slows adoption, and hinders business impact. Viewing AI holistically encourages better planning, more strategic investments, and smoother cross-department collaboration.

Moreover, this framework helps business leaders avoid a common pitfall: focusing too heavily on algorithms while underestimating the importance of data and infrastructure. In reality, a majority of AI failures occur not because the algorithm is flawed, but because the data or computational foundation is inadequate. By treating all three components as equal pillars, business organizations can build stronger, more resilient AI solutions that stand up to real-world complexity.

Ultimately, recognizing AI as an assembly of interdependent components empowers businesses to design systems that are robust, scalable, and aligned with their strategic needs. In sum, when data, algorithms, and computation work in harmony, AI becomes a reliable engine for efficiency, insight, and innovation - one capable of generating sustained business value rather than isolated experimental success.

The AI Business Use Case: Where Value is Created

Building a successful business use case for AI begins with understanding the fundamental ways it can create measurable value. While AI is a versatile technology, nearly every high-impact application falls into one of three core value levers: cost reduction, revenue generation, or risk mitigation. These levers give business organizations a clear framework for identifying where AI can make the greatest difference and for prioritizing investments that align with strategic goals. By framing AI initiatives around these outcomes, business leaders can ensure that projects stay focused, practical, and directly tied to business performance.

The first major value lever is *cost reduction through automation*, an area where AI has already proven its transformative power. AI excels at handling repetitive, rules-based, and high-volume tasks - exactly the kind of work that strains human teams and introduces operational inefficiencies. Technologies like Robotic Process Automation (RPA) and machine learning can automatically process invoices, reconcile financial statements, route documents, and validate entries with speed and precision. In financial departments, for example, automated invoice processing can dramatically reduce labor costs while minimizing errors, freeing employees to focus on more strategic work. Similar benefits appear across manufacturing, supply chain logistics, and back-office operations, where automation cuts waste, improves consistency, and accelerates turnaround times.

The second value lever centers on *revenue generation through enhancement.* Rather than replacing existing processes, AI can elevate products and services by making them smarter, more personalized, and better aligned with customer needs. In marketing and sales, AI-driven recommendation engines analyze browsing behavior, purchase history, and contextual signals to suggest items a customer is most likely to buy… significantly increasing conversion rates and average order value. Forecasting algorithms help businesses anticipate demand with greater accuracy, reducing stockouts and overproduction. Dynamic pricing systems adjust prices in real time based on market conditions, competition, and consumer behavior. These enhancements do not simply streamline operations; they actively drive top-line growth by improving customer experiences, expanding product relevance, and unlocking new revenue opportunities.

A third (and increasingly critical) value lever involves *risk mitigation and decision augmentation.* In environments where decisions must be made quickly and accurately, AI's ability to process massive datasets and detect subtle anomalies offers a decisive advantage. Consider the example of a bank analyzing billions of transactions in real time. Machine learning models can identify patterns indicative of fraud far more efficiently than traditional rule-based systems, reducing financial losses and improving customer trust. In industries like cybersecurity, predictive maintenance, compliance, and financial trading, AI systems help organizations anticipate problems before they escalate. These tools do not replace human judgment; they strengthen it by providing sharper insights, early warnings, and evidence-based recommendations.

While each value lever delivers benefits on its own, the most impactful AI initiatives often combine two or even all three levers. For instance, an AI-powered customer service system might reduce labor costs by automating routine inquiries (cost reduction), while simultaneously improving customer satisfaction through faster response times (revenue enhancement), and flagging unusual account activity (risk mitigation). These multi-lever systems magnify the return on investment and accelerate adoption across the business organization.

Understanding these value levers helps business leaders evaluate potential AI use cases and opportunities with clarity and discipline. Instead of pursuing technology for its novelty, businesses can focus on the areas where AI is most likely to deliver measurable results. This approach ensures that AI initiatives are grounded in real business needs, enabling organizations to scale their impact and build a portfolio of projects that drive sustainable competitive advantage.

Introduction to the AI Technology Stack

Artificial Intelligence systems are among the most *architecturally complex* technologies that business organizations deploy. Unlike conventional software applications, which execute predetermined logic against structured inputs, AI systems are dynamic, data-dependent, and continuously evolving artifacts that integrate hardware, data infrastructure, algorithmic models, business applications, and organizational oversight into a single operational system. Understanding this complexity is not merely a technical exercise - it represents a strategic imperative for business and technology leaders who must make consequential decisions about AI investment, deployment, and governance without necessarily possessing deep technical expertise themselves. The AI technology stack provides the conceptual framework that makes this understanding accessible, organizing the constituent components of any enterprise AI system into five distinct but interdependent layers that together describe how AI is provisioned, prepared, built, used, and governed at scale.

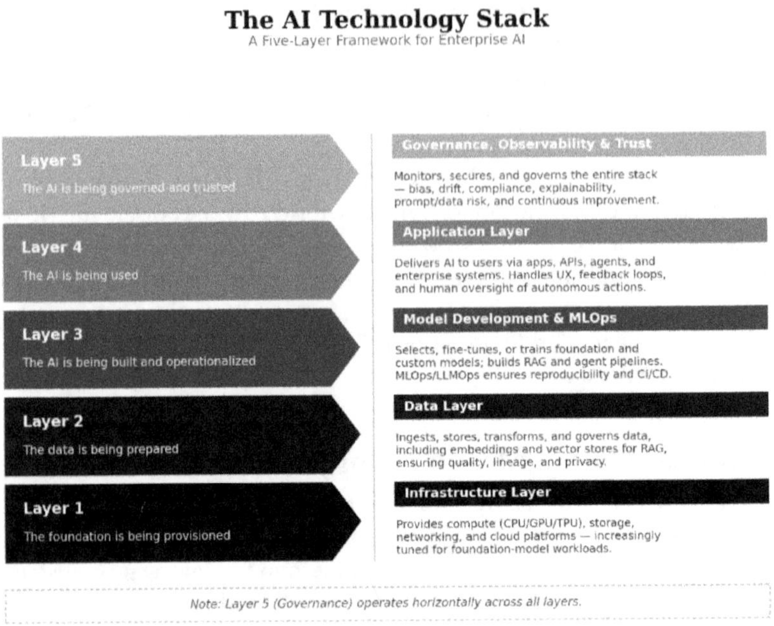

Figure 1-1 Five-Layer AI Technology Stack

The five-layer model presented in Figure 1-1 follows the natural lifecycle of an AI system from its most foundational technical requirements to its highest-level organizational responsibilities. At the bottom of the stack, raw computational capacity is provisioned to support the processing demands that AI workloads impose. Moving upward, data is collected, cleaned, and structured into the form that AI models require. The third layer is where models are developed, trained, validated, and operationalized for production deployment. The fourth layer is where those models are integrated into applications and business processes that deliver value to users and organizations. And the fifth layer, which operates horizontally across all the others, ensures that the entire system is monitored, governed, secured, and held accountable to ethical and regulatory standards over time. Taken together, these five layers capture a progression from the physical to the organizational, from the technical to the strategic, and from capability to responsibility. Now let's examine each layer in detail.

Layer 1: The Infrastructure Layer — The Foundation Is Being Provisioned

Every AI system, regardless of its sophistication or business purpose, begins with a physical and virtual foundation of computational resources. The *Infrastructure Layer* is that foundation - the hardware, storage, networking, and cloud services that make it possible to process the enormous volumes of data and perform the intensive mathematical computations that AI training and inference require. Without a well-designed infrastructure layer, no amount of algorithmic sophistication or data quality can produce a functional AI system, because the raw capacity required to execute AI workloads simply will not exist.

The core components of the infrastructure layer include compute resources, storage systems, networking capabilities, cloud platforms, and containerization and orchestration tools. Compute resources such as the processing units that perform the mathematical operations underlying AI have evolved considerably as AI workloads have grown in scale and complexity. Traditional central processing units, or CPUs, remain valuable for general-purpose computation, but the parallel processing architectures of graphics processing units, or GPUs, have become the workhorse of modern AI training because their ability to perform many calculations simultaneously maps naturally onto the matrix operations that neural network training requires. Tensor processing units, or TPUs, represent a further specialization, designed specifically to accelerate the linear algebra computations at the heart of deep learning. The selection of the appropriate compute architecture for a given AI workload is itself a consequential business decision, carrying significant implications for training speed, cost, and energy consumption.

Storage systems in the infrastructure layer must be capable of managing the scale of data that AI systems consume. Data lakes involve large repositories that store raw data in its native format, structured or unstructured, until it is needed and have become a standard component of enterprise AI infrastructure because they provide the flexibility to ingest diverse data types without requiring upfront schema definition. Distributed storage architectures, which spread data across multiple physical or virtual storage nodes, ensure that storage capacity can grow with data volumes and that no single point of failure can render the data inaccessible. Networking capabilities such as bandwidth, latency, and reliability of the connections between storage, compute, and application components are equally critical, because AI systems that depend on large-scale data transfer between components are profoundly sensitive to network bottlenecks.

Cloud computing platforms have become the dominant infrastructure environment for enterprise AI because they provide on-demand access to virtually unlimited compute and storage capacity without requiring organizations to build and maintain their own physical data centers. Public cloud providers, such as Amazon Web Services, Microsoft Azure, and Google Cloud, offer managed AI infrastructure services that abstract the complexity of hardware provisioning, allowing data science and engineering teams to focus on model development rather than infrastructure management. Private cloud and hybrid cloud configurations, which combine on-premises infrastructure with public cloud services, are increasingly common in industries with strict data sovereignty or latency requirements.

Containerization and orchestration technologies complete the infrastructure layer by providing the runtime environments in which AI components execute. *Containers* or lightweight, portable software packages that encapsulate an application and its dependencies ensure that AI models and pipelines behave consistently across different infrastructure environments, eliminating the notorious problem of software that works in development but fails in production. *Kubernetes*, the dominant container orchestration platform, automates the deployment, scaling, and management of containerized workloads, allowing AI infrastructure to scale elastically in response to demand without manual intervention.

The key considerations governing infrastructure layer design include scalability, performance, cost, and reliability. Scalability requires that the infrastructure can accommodate growth in data volumes, model complexity, and user demand without architectural redesign. Performance considerations (particularly latency, which measures the time between a request and a response, and throughput, which measures the volume of work the system can process per unit of time) directly determine the responsiveness of AI-powered applications and the feasibility of real-time use cases. Cost optimization is a persistent infrastructure challenge because AI training workloads can consume enormous computational resources over extended periods, and the economics of cloud infrastructure mean that poorly managed compute and storage costs can erode the ROI of AI investments substantially. Reliability and fault tolerance or the ability of the infrastructure to continue operating correctly in the presence of hardware failures, network disruptions, or software errors are prerequisites for deploying AI in production environments where downtime carries real business consequences.

The infrastructure layer's role in the overall stack is fundamentally enabling and does not produce intelligence, business value, or governance. Those elements are the responsibilities of the higher layers. However, this layer does ensure that sufficient computational capacity and infrastructure services are available to support everything that sits above it. An organization that underinvests in its infrastructure layer will find that even the most sophisticated models and the most carefully curated data cannot produce reliable, scalable AI systems, because the physical and virtual foundation required to execute them is inadequate.

Layer 2: The Data Layer — The Data Is Being Prepared

If the infrastructure layer serves as the foundation on which AI systems run, the data layer functions as the substance from which they learn. Machine learning models, including the large language models and deep neural networks that power the most visible AI applications of the current era, do not reason from first principles or apply human-encoded rules. They identify trends, patterns, and behaviors in data, and the quality, diversity, relevance, and completeness of that data determines, more than any other single factor, the quality of the intelligence the model will produce. The data layer encompasses all processes, systems, and practices through which raw data is collected, stored, transformed, annotated, and made ready for use in AI model training and inference.

Data ingestion is the entry point of the data layer or the process by which data from diverse source systems is brought into the AI infrastructure for processing. Ingestion pipelines, which are the automated workflows that handle this transfer, operate in two primary modes. Batch ingestion collects and transfers data in discrete, scheduled intervals, for example nightly database exports, weekly file transfers, and similar periodic processes that are appropriate when the freshness of data is not time-critical. Streaming ingestion, by contrast, captures and processes data continuously as it is generated, enabling AI systems to operate on data that is current to within seconds or milliseconds of the underlying events that produced it. As AI use cases increasingly require real-time responsiveness for applications such as fraud detection, dynamic pricing, personalized recommendation, streaming ingestion has grown from a specialized capability to a standard architectural requirement.

Data storage in the AI context has evolved considerably beyond the relational databases that served enterprise transaction and reporting needs for decades. Data warehouses, designed as structured repositories optimized for analytical queries against historical data, remain valuable for business intelligence workloads but are poorly suited to the unstructured and semi-structured data types on which AI systems most frequently depend. Data lakes, which store data in raw format across distributed storage infrastructure, offer the flexibility to accommodate diverse data modalities such as text, images, audio, video, sensor output that structured warehouses cannot manage efficiently. The data "*lakehouse*" architecture, which combines the flexible storage of data lakes with the transactional reliability and query performance of data warehouses, has emerged as the leading pattern for enterprise AI data infrastructure because it allows organizations to manage all their data, both structured and unstructured, historical and real-time, within a unified storage layer.

Data preprocessing and transformation are the activities through which raw ingested data is cleaned, normalized, and restructured into the formats that AI model training requires. Real-world data is rarely clean, complete, or consistently formatted when it arrives in the data layer. Records may be missing values, contain errors, use inconsistent naming conventions, or reflect historical biases in how data was collected. Preprocessing addresses these problems through a combination of automated validation rules, imputation techniques for handling missing values, normalization procedures that bring data from different sources onto comparable scales, and deduplication processes that eliminate redundant records. The quality of this preprocessing work has a direct and significant impact on model performance, an important principle captured in the familiar observation that AI systems are only as good as the data on which they are trained.

Data annotation and labeling deserve particular attention because they represent the human-intensive process through which raw data is assigned the labels, categories, or attributes that supervised learning algorithms require to learn from examples. An *image recognition model* cannot learn to distinguish between a defective and a non-defective manufactured component without a training dataset in which each image has been labeled by a human expert as defective or non-defective. A *natural language processing model* cannot learn to classify customer sentiment without a training corpus in which human annotators have labeled each text sample as positive, negative, or neutral. The quality, consistency, and representativeness of annotation work, which is governed by annotation guidelines, inter-annotator agreement measurement, and quality review processes, is therefore a critical determinant of model quality, and organizations that underinvest in annotation quality control consistently find that their model performance problems trace back to labeling errors and inconsistencies in the training data.

Metadata management and data catalogs provide the organizational infrastructure that makes large-scale data environments navigable and governable. A data catalog is a system of record that documents what data assets the organization possesses, where they reside, how they were collected, what they contain, who is responsible for them, how they may be used, and what transformations they have undergone. Without robust metadata management, data environments quickly become opaque. Data scientists often spend disproportionate time searching for usable datasets rather than building models and organizations cannot demonstrate to regulators or auditors how specific data was used in specific AI systems.

The data layer also encompasses critical security and compliance responsibilities that are not optional extras but foundational requirements in any enterprise AI environment. Data encryption (both at rest when data is stored and in transit, when data is moving between systems) protects sensitive information from unauthorized access. Access controls, implemented through role-based permissions and identity management systems, ensure that only authorized users and systems can read, modify, or delete data assets. Privacy protection techniques, including data masking, which replaces sensitive values with anonymized placeholders, and differential privacy, which introduces mathematically calibrated noise into datasets to prevent the identification of individual records, allow organizations to use sensitive data for AI training while protecting the privacy of the individuals whose information it contains. Regulatory compliance at the data level requires that data handling practices conform to the requirements of applicable legal frameworks, including the European Union's General Data Protection Regulation, the California Consumer Privacy Act, the Health Insurance Portability and Accountability Act for healthcare data, and the rapidly expanding body of AI-specific regulation that is emerging across jurisdictions globally.

The key considerations governing data layer design reflect the fundamental truth that data quality is not a technical nicety but a business-critical determinant of AI system value. Data quality, meaning accuracy, completeness, timeliness, consistency, and representativeness of the data, directly determines whether AI models can learn reliable patterns or will instead learn to replicate the errors, gaps, and biases present in the training data. Data lineage and traceability or the ability to track the origin and transformation history of every data element in the system are essential for auditing model behavior, diagnosing performance problems, and demonstrating regulatory compliance. Bias in training data, which occurs when the data systematically underrepresents or misrepresents certain populations or circumstances, is among the most consequential and most frequently underestimated risks in AI development, because models trained on biased data will reproduce and potentially *amplify* those biases in their predictions. And governance of sensitive data through ensuring that personally identifiable information, proprietary business data, and other sensitive assets are handled in accordance with legal requirements and organizational policies throughout the data lifecycle is a responsibility that touches the data layer more directly than any other layer in the stack.

Layer 3: The Model Development and MLOps Layer — The AI Is Being Built and Operationalized

The model development and MLOps layer is where the intelligence of an AI system is created; this layer is where data is transformed into trained algorithms capable of recognizing patterns, making predictions, generating outputs, and supporting decisions. Layer 3 encompasses the full spectrum of activities through which a business problem is translated into a functioning model: 1) selecting the appropriate algorithmic approach; 2) engineering the features from which the model will learn; 3) training the model on prepared data; 4) validating its performance against held-out test data; and 4) operationalizing it for reliable, reproducible deployment in production environments. The combination of data science practice with the operational discipline of *Machine Learning Operations*, or MLOps, is what distinguishes mature enterprise AI development from experimental prototyping.

Model selection refers to the process of identifying the algorithmic approach most appropriate to the problem at hand. The landscape of available model architectures is broad and continues to expand rapidly. *Classical machine learning approaches*, including linear and logistic regression for prediction and classification, decision trees and random forests for interpretable classification, and gradient boosting algorithms for high-performance tabular data problems, all remain widely used in enterprise settings because of their interpretability, computational efficiency, and well-understood performance characteristics. *Deep learning architectures*, namely multi-layer neural networks capable of learning complex representations from large datasets, have become the dominant approach for perceptual tasks involving images, audio, and natural language, where the complexity of the input data exceeds the capacity of classical algorithms to capture. *Transformer architectures*, which underlie the large language models that have attracted enormous public and commercial attention in recent years, have proven capable of generalizing across an extraordinary range of language understanding and generation tasks and are increasingly applied to domains beyond natural language, including biological sequence analysis, code generation, and multimodal reasoning.

The selection of a model architecture is not purely a technical decision but carries significant business implications. More complex models are generally more capable but also more expensive to train, more difficult to interpret, and more demanding of infrastructure resources for inference. In many enterprise contexts, a simpler, more interpretable model that performs adequately on the business problem is preferable to a more powerful model whose behavior cannot be explained to business stakeholders, audited by compliance teams, or scrutinized by regulators. The choice between building a model from scratch, adapting a pre-trained model through fine-tuning, or using a foundation model through prompt engineering and retrieval augmentation represents one of the most consequential architectural decisions in modern AI development, with direct implications for development cost, deployment speed, intellectual property ownership, and model performance.

Pre-trained models and transfer learning have fundamentally changed the economics of AI model development. Rather than training large models from random initialization on proprietary datasets (a process that requires enormous computational resources and large volumes of labeled training data) organizations can now begin with models that have already been trained on massive general-purpose datasets and adapt them to specific business tasks through fine-tuning on much smaller domain-specific datasets. This transfer of learned representations from a general domain to a specific application dramatically reduces the data, compute, and time required to achieve competitive model performance, making sophisticated AI capabilities accessible to organizations that could not justify the investment in training large models from scratch.

Feature engineering or the process of transforming raw data inputs into the mathematical representations, from which model training algorithms can learn, remains one of the highest-value and most intellectually demanding activities in the model development layer. Even as deep learning architectures have automated feature extraction for perceptual data like images and text, feature engineering retains enormous importance for tabular business data, where the selection, construction, and transformation of input features often determine the ceiling of model performance more decisively than the choice of algorithm. The calculation of derived features (e.g., ratios, aggregations, time-based statistics, interaction terms, and domain-specific indicators) requires deep understanding of both the business domain and the statistical properties of the data, combining the knowledge of subject matter experts with the technical skill of data scientists.

Model training is the computationally intensive process through which the selected algorithm adjusts its internal parameters to minimize prediction error on the training dataset. The configuration of training, including the selection of loss functions that define what error means for the problem at hand, optimization algorithms that determine how parameters are updated, and hyperparameters that govern the training process itself, requires iterative experimentation and systematic comparison of results. Hyperparameter tuning, the process of searching the space of possible training configurations to identify the combination that produces the best-performing model on validation data, has been increasingly automated through techniques such as grid search, random search, and Bayesian optimization, reducing the manual effort required while expanding the breadth of the configuration space that can be explored.

Validation and testing ensure that the model's performance on training data generalizes to new, unseen data, a fundamental requirement for a model to be useful in the real world. *Overfitting*, which occurs when a model learns the specific patterns of the training data so precisely that it fails to generalize to new examples, is among the most common failure modes in model development and one that thorough validation procedures are designed to detect. *Underfitting*, which occurs when the model is insufficiently complex or insufficiently trained to capture the genuine patterns in the data, is equally problematic but typically easier to detect because it manifests as poor performance even on the training data. The design of train-validation-test splits, cross-validation schemes, and held-out evaluation datasets, along with the rigorous application of these evaluation practices throughout model development is what separates reliable model performance assessment from the optimistic but misleading performance numbers that undisciplined evaluation procedures generally produce.

MLOps capabilities transform the model development process from a research activity into an operational discipline. Experiment tracking systems record the configuration, training data, hyperparameters, and performance metrics of every model training run, creating a searchable history of experimentation that enables systematic comparison of approaches and prevents the common problem of valuable experiments being lost or forgotten because their results were not systematically recorded. Model versioning ensures that every trained model, along with the data and code that produced it, is preserved in a form that can be retrieved, reproduced, and audited. This requirement is becoming increasingly important as regulatory frameworks begin to demand documentation of the training lineage of AI systems deployed in consequential contexts. Continuous integration and continuous deployment pipelines for models automate the testing, validation, and deployment of new model versions, applying to AI systems the same software engineering rigor that DevOps practices have established for conventional software. Model packaging and deployment readiness or the preparation of trained models for serving in production environments, including format conversion, optimization for inference hardware, and integration with model serving infrastructure bridges the gap between the research environment in which models are trained and the production environment in which they must operate reliably at scale.

The key considerations governing the model development and MLOps layer reflect the dual nature of its responsibilities. On the model quality side, accuracy and generalization or rather the model's ability to perform well on new examples drawn from the same distribution as its training data are the primary performance criteria, supplemented by fairness metrics that assess whether performance is consistent across demographic groups and use cases, and interpretability metrics that assess whether model behavior can be explained in terms that business stakeholders and regulators can evaluate. On the operational side, training efficiency and cost such as the computational resources consumed and the time required to train models to production-ready performance will determine the economics of AI development and the feasibility of frequent model updates. Reproducibility and auditability, basically the ability to recreate a specific model training run and demonstrate exactly how a deployed model was produced, are not merely engineering best practices but increasingly legal and regulatory requirements that enterprise AI programs must satisfy.

Layer 4: The Application Layer — The AI Is Being Used

The application layer is where artificial intelligence crosses the threshold from technical capability to business value. Layer 4 is the layer at which trained models are integrated into the products, services, and workflows with which users interact and the point where the intelligence created in Layer 3, trained on the data prepared in Layer 2, running on the infrastructure provisioned in Layer 1, becomes visible and useful to the employees, customers, and partners whose work and decisions the AI system is designed to support. Without a well-designed application layer, even the most capable AI models remain inaccessible and will stay confined to research environments and unable to influence the business outcomes that initially justified the AI investment.

The user-facing applications that constitute the visible face of the application layer take many forms across enterprise contexts. Web and mobile applications provide direct access to AI capabilities through consumer and employee interfaces, for example the customer service chatbot that resolves inquiries without human intervention, the recommendation engine that personalizes product displays, or the predictive analytics dashboard that surfaces demand forecasts for operations managers. Enterprise systems, including customer relationship management (CRM) platforms, enterprise resource planning (ERP) systems, electronic health records, and the full range of systems-of-record that manage the core data and processes of the organization are increasingly embedding AI capabilities directly into their interfaces and workflows, often through the AI features that enterprise software vendors are rapidly integrating into their platforms. These embedded AI capabilities are becoming an important channel for AI value delivery because they meet users where they already work, eliminating the adoption friction that standalone AI tools frequently encounter.

Application programming interfaces (APIs) represent the technical mechanisms through which AI model capabilities are made accessible to application developers and integrated into existing systems. A trained model deployed behind an API can receive a request from any authorized application, perform inference, and return a prediction or generated output without the calling application needing to know anything about the model's internal architecture or training. This decoupling of model capability from application logic is architecturally important because it allows models to be updated, replaced, or scaled independently of the applications that consume them, and because it enables the same model capability to be consumed by multiple applications simultaneously. Microservices architectures, which decompose complex applications into small, independently deployable services that communicate through APIs, have become the dominant pattern for enterprise application development and are well-suited to AI integration because they allow AI capabilities to be added, modified, or removed without requiring changes to unrelated application components.

Business logic and workflow orchestration bridge the gap between AI model outputs such as predictions, scores, classifications, or generated content and the business decisions and actions they are intended to support. AI models *do not* make business decisions; they produce outputs that inform business decisions, which are then acted upon through the processes, rules, and workflows that govern how the organization operates. The application layer is where these connections are forged, where a fraud detection model's risk score is translated into a transaction approval or rejection decision, where a demand forecasting model's output is translated into a replenishment order, or where a document understanding model's extraction is translated into a populated form in a downstream system. The design of this business logic layer, essentially determining when to act on AI outputs automatically and when to route them for human review, how to handle model uncertainty, and how to integrate AI recommendations into existing approval workflows is as important to AI value delivery as the design of the model itself.

User interface and experience design for AI-powered applications requires attention to dimensions that conventional application design does not address. Users interacting with AI systems must understand what the system can and cannot do, must have appropriate expectations about the reliability and limitations of AI outputs, and must have clear pathways to override or escalate AI recommendations when their judgment or contextual knowledge indicates that the model's output is wrong or inappropriate. The design of explanations (which communicate why the AI system produced a particular output in terms that users can evaluate and act on) is a discipline that sits at the intersection of machine learning and human-computer interaction and is *increasingly recognized* as a prerequisite for user trust and adoption. Personalization and context management, which allow AI applications to tailor their behavior to the specific user, session, and situational context in which they are operating, are increasingly standard requirements for user-facing AI applications across consumer and enterprise contexts.

The security and access control responsibilities of the application layer focus on ensuring that AI capabilities are accessible only to authorized users and systems and that the interactions users have with AI systems are appropriately authenticated, authorized, and logged. Identity and access management systems are those which govern who can access what resources under what conditions must be extended to cover AI model endpoints and the sensitive outputs they produce. Role-based access control, which grants different levels of AI system access based on the user's organizational role, is a standard mechanism for managing this authorization complexity at enterprise scale. Secure API access results in ensuring that model-serving endpoints are protected against unauthorized access, injection attacks, and data exfiltration is a security requirement that applies to every AI model deployed through an API interface.

Feedback and interaction mechanisms are among the most strategically valuable and most frequently under-designed components of the application layer. User feedback, especially the explicit signals that users provide when they rate a recommendation, correct an AI-generated output, or report an inaccurate prediction is among the richest training signal available for model improvement, because it reflects the judgment of domain experts operating in real business contexts. Human-in-the-loop processes, which route AI outputs to human reviewers for validation or correction before they are acted upon, serve both a quality assurance function (e.g., preventing consequential errors from propagating unchecked) and a continuous improvement function because the corrections that human reviewers make to AI outputs provide labeled training examples that can be used to improve future model performance. Interaction logging, which records the inputs, outputs, and contextual metadata of every user interaction with an AI system, creates the audit trail that is essential for diagnosing model failures, demonstrating regulatory compliance, and understanding how AI capabilities are being used in practice today.

The key considerations governing application layer design reflect the reality that this is the layer where AI systems meet organizational reality. Usability and user trust determine whether AI capabilities are adopted and used effectively or ignored and circumvented. This distinction has enormous implications for AI ROI. System responsiveness and latency determine whether AI-powered experiences feel seamless and natural or frustrating and slow, a consideration that is particularly acute for real-time use cases where users have no tolerance for delays. Integration complexity, meaning the technical and organizational challenges of connecting AI capabilities to the existing systems, data flows, and processes of the organization, is frequently the most significant practical obstacle to AI value delivery, primarily because enterprises have accumulated decades of legacy systems that were not designed to interoperate with AI components. Also, alignment with business processes (meaning the degree to which AI capabilities are embedded into the workflows and decision-making processes that govern how work gets done) is the ultimate determinant of whether AI investments generate business outcomes or merely generate impressive demonstrations.

Layer 5: The Governance, Observability, and Trust Layer: The AI Is Being Watched, Controlled, and Held Accountable

The fifth layer of the AI technology stack is unique in its scope and character. While the first four layers are sequential with each enabling the next, the governance, observability, and trust layer is horizontal, spanning the entire stack and exercising oversight across every component in every other layer simultaneously. Layer 5 is the layer that transforms AI from a technical system into an enterprise system and one that can be monitored for performance degradation, audited for compliance, defended against security threats, evaluated for fairness, and held accountable for its impacts on users, organizations, and society. As AI systems take on increasingly consequential roles in enterprise decision-making, the governance layer has shifted from an afterthought appended to deployed systems to a foundational requirement that must be designed into AI architecture from the beginning.

Observability or the ability to understand the internal state of a system by examining its external outputs is the first major responsibility of the governance layer. In the AI context, observability encompasses model performance monitoring, data drift detection, logging, and alerting. Model performance monitoring tracks the accuracy, precision, recall, and other quality metrics of deployed models over time, detecting the gradual degradation that occurs as the statistical properties of the real-world data the model encounters diverge from the properties of the data on which it was trained. This phenomenon, known as *model drift* or *concept drift*, is an inherent characteristic of AI systems deployed in dynamic environments, and the failure to monitor for it is a leading cause of the slow-motion model failures that damage business outcomes and erode user trust without producing the clear, immediate error signals that conventional software failures generate!

Data drift detection, which monitors changes in the statistical distribution of the inputs that the model receives rather than the quality of its outputs, is an *early warning system* for the model performance degradation that will eventually follow if input distributions shift significantly from training distributions. For example, a fraud detection model trained primarily on transaction data from pre-pandemic consumer behavior may begin to perform poorly as consumer spending patterns shift - not because the model has changed but because the world it is operating in has changed in ways that invalidate the patterns it learned. Data drift monitoring can detect these distributional shifts early, before they have fully propagated into model performance degradation, providing the signal that triggers investigation and, when warranted, model retraining. Logging and traceability across pipelines, including the systematic recording of what data flowed through which system components at what time and produced what outputs, creates the audit trail that makes it possible to diagnose model failures, investigate anomalous outputs, and demonstrate to regulators exactly how a specific AI decision was produced.

Governance in the fifth layer addresses the organizational and procedural controls that ensure AI systems are developed, deployed, and operated in accordance with the organization's policies, values, and regulatory obligations. Model lifecycle management involves the systematic tracking of model versions, training lineage, deployment history, and retirement decisions and is the organizational foundation of AI governance, creating the institutional memory that allows organizations to understand and account for the behavior of every model currently in production. Approval workflows and deployment controls formalize the process by which new models and model updates are reviewed, tested, and authorized for production deployment, ensuring that consequential AI systems are subject to appropriate scrutiny before they begin influencing real decisions. Audit trails and documentation involving the comprehensive records of how models were developed, what data trained them, what performance standards they were validated against, and what deployment decisions were made about them are increasingly required by regulators and are essential for the internal accountability that responsible AI governance demands.

Security in the governance layer extends the security considerations that apply individually to the data layer and the application layer into a comprehensive, system-wide security posture that addresses the unique vulnerabilities of AI systems. Data and model protection includes ensuring that training datasets, model weights, and inference outputs are accessible only to authorized parties and is important both for protecting proprietary AI investments and for preventing the leakage of sensitive information that training data may contain. API and system security is concerned with hardening the interfaces through which AI systems communicate with the applications and data sources they interact with and is critical because AI model-serving endpoints represent attractive attack surfaces for adversaries seeking to extract model predictions, manipulate model behavior, or exfiltrate training data through model inversion attacks. Defense against adversarial attacks such as the deliberate manipulation of AI model inputs to cause incorrect or harmful outputs is a security discipline specific to AI systems that has no precise analogue in conventional cybersecurity. Adversarial examples, which are inputs crafted to cause model misclassification through imperceptible modifications that humans cannot detect, represent a qualitatively different category of security threat from the injection attacks and authentication bypasses that conventional application security addresses.

The trust and responsible AI components of the governance layer address the ethical dimensions of AI system behavior. These dimensions reflect the qualities that determine whether AI systems are fair, transparent, explainable, and aligned with human values. Bias detection and mitigation are ongoing governance responsibilities and not simply one-time activities performed before deployment. Models that demonstrate equitable performance across demographic groups at the time of deployment can become biased over time as the distribution of users and inputs changes, or as social conditions shift in ways that alter the relationship between model inputs and the outcomes the model predicts. Continuous bias monitoring or tracking fairness metrics across protected categories throughout the deployment lifecycle is therefore essential for organizations committed to responsible AI operation. Explainability and interpretability or the ability to account for why a model produced a specific output in terms that users, business stakeholders, and regulators can understand are increasingly required both by emerging regulatory frameworks and by the practical demands of human-AI collaboration, where users who cannot understand or evaluate AI recommendations are poorly positioned to exercise the human oversight that responsible deployment requires.

Compliance and regulatory alignment in the governance layer ensures that AI systems meet the requirements of the legal and industry-specific frameworks that govern their deployment. The National Institute of Standards and Technology (NIST) AI Risk Management Framework provides a comprehensive, voluntary framework for identifying, assessing, and managing the risks that AI systems present, organized around the functions of governing, mapping, measuring, and managing AI risk. International Standards Organization (ISO) standards for AI systems (including ISO 42001 which establishes requirements for AI management systems) provide internationally recognized benchmarks against which organizations can assess and certify their AI governance practices. Industry-specific regulations, including the requirements of the European Union's AI Act, which mandates conformity assessments and transparency obligations for high-risk AI applications, Health Insurance Portability and Accountability Act (HIPAA) requirements for AI systems that process protected health information, and the financial services regulations that govern AI-assisted credit decisioning impose binding compliance obligations that the governance layer must be designed to satisfy. Organizations that treat regulatory compliance as a retroactive compliance exercise rather than a design requirement consistently find that achieving compliance after deployment is far more expensive and disruptive than building compliant systems from the beginning!

The continuous improvement mechanisms of the governance layer complete the feedback loop that allows AI systems to evolve and improve over their operational lifetimes. Feedback loops for retraining, especially the processes by which user corrections, model performance data, and new training examples are incorporated into the datasets used to train updated model versions, are what allow AI systems to improve with use rather than degrading over time as their training data grows stale. A/B testing and model comparison or the systematic evaluation of new model versions against existing ones in controlled production experiments provide the empirical evidence that model updates in-fact do improve the metrics that matter to the business, preventing the premature deployment of model changes that look promising in evaluation but underperform in production. Performance-based retraining triggers (referring to automated systems that initiate model retraining when performance metrics fall below defined thresholds) transform the governance of model quality from a manual monitoring activity into an automated operational process, ensuring that model degradation is addressed systematically rather than reactively after business impact has already occurred.

The AI Stack as an Integrated System: The Critical Insight

Having examined each layer of the AI technology stack in detail, let's now step back to articulate the insight that the layered model is designed to convey: one which enterprise AI leaders most frequently fail to internalize until they have experienced the consequences of ignoring it. *AI systems are not defined by their models.* A model, however sophisticated, is a single component in a five-layer system, and the performance, reliability, fairness, and business value of the overall system depend as much on the quality of the other four layers as on the quality of the model itself.

This insight has direct and practical implications for how organizations should think about AI failure. When an AI system fails to deliver expected business value (a circumstance that is, as other chapters of this book discuss in detail, more common than the AI industry's marketing materials would suggest) the root cause almost always lies not in the model but in one of the surrounding layers. For example, poor data quality in Layer 2 produces models that have learned to replicate errors and biases in the training data rather than genuine patterns. Inadequate data annotation produces models that have learned the wrong labels. Weak deployment design in Layer 4 produces models that cannot be accessed by the users who need them, cannot be integrated into the business workflows they are designed to support, or cannot scale to the volume of requests the production environment demands. Lack of governance and monitoring in Layer 5 produces models that degrade in silence as the world changes around them, generating increasingly unreliable outputs long after they should have been retrained or retired.

The layered model also clarifies where competitive AI advantage is most durably created. While the research and business press tends to focus on model capability or the benchmarks that large language models achieve, the parameter counts that distinguish one foundation model from another, the most defensible sources of enterprise AI advantage lie in the data layer, where proprietary datasets accumulated over years of operation create training advantages that competitors cannot easily replicate, and in the governance layer, where the organizational disciplines of monitoring, evaluation, and continuous improvement allow organizations to sustain and compound their AI capabilities over time while competitors' systems degrade. Building and governing these capabilities effectively requires treating the AI technology stack not as a collection of independent technical components but as an *integrated system* in which every layer's quality affects every other layer's performance.

General Conclusion for Business

The five-layer AI technology stack (Infrastructure, Data, Model Development and MLOps, Application, and Governance) provides a comprehensive and practically useful framework for understanding how AI systems are built, deployed, and sustained in enterprise environments. For *business leaders*, the value of this framework lies not in the technical details of any individual component but in the structured vocabulary it provides for asking the right questions about AI investments: Is our infrastructure layer capable of supporting the AI workloads we are planning? Is our data layer producing the quality and diversity of data that our models require? Is our model development and MLOps layer applying the operational discipline that production-grade AI deployment demands? Is our application layer successfully embedding AI capabilities into the workflows and decisions where they create business value? Is our governance layer providing the monitoring, accountability, and continuous improvement mechanisms that ensure our AI systems remain reliable, fair, and compliant over time?

As AI adoption matures and competitive pressure to deploy AI at scale intensifies, the organizations that achieve durable advantage will not necessarily be those that build the most sophisticated models. They will be those that most effectively integrate and govern AI capabilities across the entire stack by managing data as a strategic asset, operationalizing model development as a repeatable discipline, embedding AI into business applications with the care that user trust requires, and maintaining the governance posture that responsible enterprise-scale AI deployment demands. The AI technology stack is not merely a technical architecture, but rather the organizational blueprint for turning artificial intelligence platforms from a promising technology into a sustainable competitive capability.

AI, Automation, and Human Collaboration

A few final points to further distinguish AI from traditional automation that are essential for business executives who are planning digital transformation initiatives or assessing where to invest resources. Although the terms are sometimes used interchangeably, they represent fundamentally different capabilities. *Automation* in general refers to executing a predefined sequence of steps without deviation, as in following a script, rule, or workflow exactly as it was programmed. Examples include tasks like exporting data from a database, formatting a report, or uploading files to a server. These systems do not learn, adapt, or make decisions; they simply repeat the same actions reliably and efficiently. Automation is powerful for stabilizing routine processes, but its value is limited to predictable, rule-driven scenarios.

AI, by contrast, represents intelligent automation or expressly the ability to make decisions, adapt to new information, and handle situations that were not explicitly programmed in advance. An AI system may adjust a delivery route based on real-time traffic, detect anomalies in financial transactions it has never encountered before, or personalize product recommendations for a new customer based on subtle behavioral cues. This capacity to learn from data and operate under uncertainty is what differentiates AI from simple scripts or macros. Instead of following rigid instructions, AI models evaluate patterns, probabilities, and outcomes to determine the best course of action.

Because AI has these adaptive qualities, the managerial focus should shift away from the misguided idea of "replacing people" and toward the more productive strategy of *augmentation*. AI is at its best when it handles data-heavy, repetitive, or time-consuming tasks that burden human experts. By offloading this workload to intelligent systems, employees can spend more time on high-value activities, things like creative problem-solving, strategic planning, interpersonal communication, and judgment-based decisions that require human insight. This approach does not diminish the role of people; it amplifies their ability to perform.

A clear example of this collaborative dynamic is found in the legal profession. A lawyer does not need an AI assistant to negotiate a settlement or craft a courtroom argument. What they DO benefit from is an AI assistant capable of scanning hundreds of thousands of documents in minutes, identifying relevant clauses, and flagging potential risks. The AI dramatically speeds up the research phase, but the lawyer still applies expertise, reasoning, ethical consideration, and nuance to interpret the findings and determine the right course of action. The value comes from the partnership and not from trying to replace the human.

An augmentation mindset regarding AI is essential for maximizing return on AI investment. When organizations view AI as a tool that empowers their workforce rather than displaces it, adoption increases, resistance decreases, and collaboration deepens. Employees feel supported rather than threatened, and they are more likely to engage with AI-enabled workflows. Leaders, in turn, can design more effective change-management strategies that acknowledge the complementary strengths of humans and machines.

Ultimately, positioning AI as a collaborator rather than a competitor produces better outcomes for both people and organizations. This approach encourages smarter project scoping, fosters a healthier workplace culture, and directs investment toward solutions that genuinely enhance productivity and innovation. By focusing on augmentation, companies can unlock AI's true potential while preserving, and even elevating, the uniquely human capabilities that drive long-term success.

CHAPTER 2

Core AI Technologies Business Leaders Must Understand

Introduction: Why Technology Literacy Is a Leadership Imperative

ARTIFICIAL INTELLIGENCE HAS BECOME A central strategic capability for modern enterprises, but its successful adoption depends on business leaders who also possess a fundamental understanding of the underlying technologies that power contemporary AI systems. While business and technology executives need not become data scientists or engineers, they must understand the capabilities and limitations of the core technological pillars, namely Machine Learning, Deep Learning, Natural Language Processing, Computer Vision, and the compute infrastructure that enables them, as well as the data architecture, cost economics, and governance structures that determine whether AI investments generate durable value.

This chapter provides a managerial and conceptual overview of these foundational technologies, explaining not only how they work but why they matter for business strategy, operational transformation, and long-term competitiveness. Chapter 2 also addresses several topics that are frequently omitted from executive AI curricula but are essential for informed decision-making: 1) the organizational structures required to govern data at scale; 2) the economics of cloud and compute investment; 3) the regulatory constraints that shape infrastructure choices; and 4) the data architecture debates that determine how AI-ready an enterprise's foundations actually are. Leaders who understand these dimensions will make better investment decisions, ask better questions of their technical teams, and avoid the strategic mistakes that most commonly undermine enterprise AI programs.

Machine Learning: The Core Engine of Modern AI

Machine Learning constitutes the core engine of modern AI. At its essence, machine learning refers to computational methods that enable systems to learn trends, patterns, and behaviors from data and make predictions or decisions directly from that data, rather than following explicit, rule-based programming. Whereas traditional software relies on predefined logic (if X, then Y) machine learning infers relationships statistically, enabling flexibility, adaptation, and scale that rule-based systems cannot achieve. For business and technology leaders, understanding machine learning means understanding how data is transformed into predictive insight.

Machine learning is commonly categorized into three principal paradigms: *supervised* learning, *unsupervised* learning, and *reinforcement* learning. Each paradigm reflects a different way a system learns from data and is suited to different classes of business problems.

Supervised learning is the most widely used form of machine learning in business. The system learns from labeled historical data in which the correct output is known, effectively mapping input variables such as customer demographics, browsing behavior, or historical sales volumes to a specific output, such as churn probability or demand forecast. Because supervised learning predicts quantifiable outcomes, this paradigm is ideally suited for problems involving forecasting, risk scoring, classification, or recommendation. Industries as diverse as retail, financial services, insurance, manufacturing, and healthcare rely on supervised learning for high-stakes decision-making, from determining the likelihood of loan default to optimizing pricing strategies.

Unsupervised learning operates without labeled outputs. Instead, the algorithm identifies hidden structures, patterns, or groupings within data. This paradigm is particularly valuable for exploratory analyses where the goal is to understand natural groupings or anomalies rather than predicting a specific outcome. Customer segmentation, product clustering, fraud detection, and pattern discovery in sensor data are common applications. Unsupervised learning enables leaders to uncover latent insights, such as discovering unexpected customer micro-segments or identifying operational abnormalities, that would be difficult to detect with traditional analytics.

Reinforcement learning represents a third, distinct paradigm centered on sequential decision-making. Reinforcement learning systems learn through interaction with an environment, receiving rewards for desirable actions and penalties for undesirable ones. Over time, the system learns a policy or a strategy for choosing actions that maximize cumulative reward. This approach is particularly valuable in dynamic settings in which decisions unfold over time and must be optimized holistically. Examples include autonomous robotics, supply-chain routing, real-time bidding in advertising markets, smart energy grids, and algorithmic trading. Reinforcement learning is still emerging in enterprise settings due to its requirements for robust simulation environments and long training cycles, but its potential for automating complex operational decisions is profound.

Together, these three paradigms form the conceptual foundation of machine learning, enabling business organizations to extract value from structured and unstructured data at scale. For leaders, the critical insight is that machine learning is not a monolithic technology but a suite of approaches, each offering unique capabilities aligned to specific business objectives.

Deep Learning: Unlocking High-Dimensional Intelligence

Deep Learning has reshaped the trajectory of artificial intelligence over the past years, enabling breakthroughs in perception, language understanding, and generative modeling. Deep learning refers to machine learning models built from neural networks or layered computational structures loosely inspired by the architecture of the human brain. What distinguishes deep learning from traditional machine learning is the depth of these networks: multiple layers of interconnected nodes that progressively transform raw input into increasingly abstract representations.

The defining innovation of deep learning lies in its capacity for automatic feature extraction. Traditional machine learning required domain experts to define relevant features manually, for example, specifying which edges, textures, or word patterns were meaningful. Deep learning algorithms, by contrast, learn such features directly from the raw data. In image recognition, a deep model learns edges in early layers, textures in subsequent layers, and full objects in deeper layers, all without human intervention. This capacity is what enables computers to recognize faces, transcribe speech, translate languages, and generate photorealistic images.

Deep learning's advances have redefined performance standards across numerous industries. In healthcare, deep neural networks interpret radiology images with accuracy rivaling expert physicians. In transportation, they enable sensor fusion and perception for autonomous vehicles. In media and entertainment, neural networks power content recommendation and generative creativity. In financial services, they detect fraud patterns far too complex for rule-based systems. For business and technology leaders, the key managerial insight is that deep learning unlocks unprecedented accuracy in tasks involving high-dimensional, complex data; however, at a significant cost.

Deep learning models require substantial volumes of labeled data, specialized expertise, and powerful compute infrastructure, particularly during training. Their training cycles can extend for days or weeks, and their complexity often makes them less interpretable than traditional models. As a result, deploying deep learning at scale requires careful alignment of data strategy, infrastructure planning, governance practices, and business justification. The benefits can be transformative, but only when business organizations invest in the necessary foundations and manage the associated risks.

Natural Language Processing: Machines That Understand Language

Natural Language Processing (NLP) is the domain of AI dedicated to enabling machines to understand, interpret, generate, and interact using human language. Business leaders encounter NLP across a broad array of applications, from sentiment analysis of customer feedback to enterprise search, document processing, conversational agents, and automated content generation. Modern NLP systems can read, classify, summarize, translate, and generate complex language, transforming how organizations engage with text and speech at scale.

Early NLP applications focused on text mining, essentially extracting structured information from large volumes of unstructured text. Systems were designed to identify keywords, entities, topics, and relationships, enabling organizations to convert raw text into actionable data. Sentiment analysis extended these capabilities by determining emotional valence in social media posts, reviews, and communications, helping companies monitor brand health, customer satisfaction, and emerging risks.

The emergence of *transformer-based architectures* revolutionized the NLP landscape. Transformers introduced self-attention mechanisms that allow models to capture intricate relationships between words across long textual sequences. This advancement paved the way for large language models (LLMs), such as GPT-class models, which are pre-trained on vast corpora of text and can be fine-tuned for a wide variety of tasks. LLMs enable sophisticated generative capabilities, from drafting emails and customer communications to writing code, summarizing lengthy documents, and answering domain-specific questions using corporate knowledge bases.

For business organizations, LLMs represent not merely automation tools but cognitive amplifiers, i.e., systems capable of synthesizing complex information, enhancing human productivity, and democratizing access to institutional expertise. Yet they also raise questions of accuracy, hallucination risk, privacy, and alignment that must be managed carefully. Leaders deploying NLP must balance the efficiency gains with appropriate oversight mechanisms and risk controls, including retrieval-augmented generation (RAG) systems that ground model outputs in authoritative, organization-specific knowledge bases.

Computer Vision: Teaching Machines to See and Interpret

Computer Vision (CV) is the field of AI focused on enabling machines to interpret visual information such as images, videos, and sensor data and extract meaningful insights to support decisions. CV systems allow computers to answer questions that humans answer effortlessly: What objects are present? Where are they located? What actions are occurring? Who is in the frame? Is something abnormal happening?

Modern computer vision relies primarily on deep learning and specifically on *Convolutional Neural Networks* (CNNs). These architectures are designed to process images efficiently by identifying local patterns such as edges, corners, and textures and combining them hierarchically to recognize full objects and scenes. CNNs have enabled breakthroughs in facial recognition, object detection, image segmentation, autonomous navigation, and medical diagnostics.

Computer vision's business value spans a wide range of operational domains. Manufacturing plants rely on CV-driven quality assurance to detect microscopic defects on fast-moving assembly lines. Logistics companies use CV for package tracking, automated sorting, and load optimization. Healthcare organizations employ CV for interpreting radiology scans, identifying anomalies, and assisting diagnostic workflows. Retailers deploy CV for in-store analytics, shelf monitoring, and theft detection. In heavily regulated industries such as construction and energy, CV systems ensure compliance by monitoring safety equipment, hazardous activity, and procedural adherence.

The technical sophistication of CV continues to grow as models today become more capable of understanding complex scenes, recognizing human actions, and integrating information across modalities. The convergence of computer vision with natural language processing (seen in multimodal models that process images and text jointly) promises even more powerful capabilities, enabling automated chart reviews, video-based incident analysis, and robotics applications that require contextual understanding. For infrastructure planners, multimodal AI introduces distinct requirements: image and video data must be stored, indexed, and served at scale alongside text, requiring storage architectures that support large binary objects, specialized retrieval pipelines, and inference infrastructure capable of handling multiple data modalities simultaneously. As a result, business organizations should plan their data platform architecture to accommodate these multimodal workloads from the outset rather than retrofitting text-centric systems.

The Silicon Beneath the Intelligence: Hardware Architectures Powering Modern AI

When most people imagine artificial intelligence at work, they picture software such as algorithms, neural networks, or lines of code executing abstract logic. Far fewer consider the physical hardware on which that intelligence depends. Yet the choice of silicon is not merely an engineering footnote; it fundamentally shapes what AI systems can do, how fast they can do it, and where they can be deployed. Today, five distinct hardware architectures power the AI landscape, each representing a different philosophical trade-off between generality and specialization. Understanding these trade-offs is essential for anyone designing, deploying, or governing AI systems at enterprise scale.

The story of AI hardware is, at its core, a story of increasing specialization. The progression from CPU to LPU mirrors the broader arc of engineering maturity. As problems become well-understood, purpose-built tools consistently outperform general ones. AI's core computation (repeated matrix multiplication) is simple enough to optimize aggressively, and the performance gains are substantial.

Five Architectures Across the Generality–Specialization Spectrum

Central Processing Unit (CPU)

The CPU is the foundation of modern computing. Designed for general-purpose work, a CPU contains a relatively small number of powerful cores, each capable of handling complex logic, conditional branching, and system-level operations. This flexibility makes CPUs indispensable for operating systems, databases, and the orchestration layer of most software applications. However, that same flexibility makes CPUs poorly suited for the workloads at the heart of neural networks. Training and running AI models requires performing the same mathematical operations, particularly matrix multiplications that occur billions of times in parallel. A CPU's few cores, no matter how powerful individually, are architecturally mismatched to this kind of repetitive, parallelizable computation. CPUs remain essential in AI pipelines, but rarely as the primary compute engine.

Graphics Processing Unit (GPU)

The GPU was not originally designed for AI, yet it became AI's dominant hardware platform, a telling example of architectural fit trumping original intent. Where a CPU has a handful of powerful cores, a GPU contains thousands of smaller, simpler cores designed to execute the same instruction simultaneously across different data. This property, known as data parallelism, maps almost perfectly onto the matrix multiplications that underpin neural network computation. *GPUs dominate AI training for this reason.* When a model needs to update millions of weights based on a batch of training examples, a GPU can process the entire batch in parallel rather than sequentially, reducing training runs that would take years on a CPU to days or weeks on a GPU cluster.

Tensor Processing Unit (TPU)

Google's TPU represents the next step: a chip designed from the ground up for neural network workloads, rather than adapted from another purpose. At the heart of a TPU is a *systolic array* or a grid of multiply-accumulate (MAC) units through which data flows in a wave-like pattern. Model weights enter the array from one side while input activations enter from another, and partial results propagate through the grid without repeatedly accessing external memory. This architecture eliminates much of the memory bandwidth bottleneck that limits GPU performance on certain workloads. TPUs excel at the large, structured matrix operations that define modern deep learning and represent a philosophically important moment: purpose-built silicon explicitly shaped around the mathematical structure of neural networks.

Neural Processing Unit (NPU)

While GPUs and TPUs are designed for data center scale, NPUs address a different constraint: power consumption. An NPU is an edge-optimized processor built around a neural compute engine packed with MAC arrays and on-chip SRAM. The governing design goal is to run AI inference, i.e., applying a trained model to new inputs within a power budget of just a few watts. This design makes NPUs the architecture of smartphones, wearables, and IoT devices. Apple's Neural Engine and Intel's NPU represent prominent examples. When a phone recognizes a face, transcribes speech, or applies a computational photography effect in real time, it is almost certainly doing so on an NPU. The trade-off is one of scope: NPUs are optimized for inference on relatively compact models, not for training large ones.

Language Processing Unit (LPU)

The newest entrant to this landscape is the LPU, pioneered by Groq, an American artificial intelligence company that builds an AI accelerator application-specific integrated circuit. Its defining architectural choice is radical: all model weights are stored in on-chip SRAM, thus eliminating off-chip memory access entirely from the critical execution path. The result is fully deterministic execution with zero cache misses and no runtime scheduling overhead. The practical consequence is dramatically lower latency, particularly for inference on large language models. The trade-off is significant: on-chip SRAM is expensive and physically limited, meaning a single LPU chip can only hold a fraction of a large model's weights, requiring many chips to be linked together. For applications where response latency is the primary constraint, e.g., real-time conversational AI, the LPU's determinism and speed offer a genuine and measurable advantage.

A Decision Framework for Hardware Selection

These five architectures can be understood along a single dimension: the spectrum from *generality* to *specialization.* Each step toward specialization surrenders flexibility in exchange for efficiency. CPUs can run any computation; LPUs can run specific computations faster than almost anything else. Neither is universally superior; the right choice depends entirely on the workload, deployment environment, and performance requirements. Table 2-1 summarizes the key trade-offs for strategic planning purposes.

Table 2-1: Comparison of Hardware Architectures Powering Modern AI

Architecture	Primary Use Case	Key Advantage	Key Trade-off
CPU	Orchestration, logic, control plane	Fully general-purpose; handles any computation	Too few parallel cores for AI math at scale
GPU	Model training; large-batch inference	Massively parallel; proven at scale	High power draw; costly at inference volume
TPU	Large-scale training and inference	Memory-efficient systolic array design	Optimized for Google's software ecosystem
NPU	Edge inference (mobile, IoT, wearables)	Extreme power efficiency; on-device privacy	Limited to compact models; no training workloads

Architecture	Primary Use Case	Key Advantage	Key Trade-off
LPU	Low-latency LLM inference	Deterministic, ultra-low latency responses	On-chip memory limits model size per chip

Hardware architecture is not an academic concern for AI practitioners but a practical design input. The architecture underlying a deployment affects latency, throughput, cost, and the physical contexts in which a system can operate. As AI systems grow more sophisticated and more embedded in everyday life, the ability to reason about hardware (not just software) becomes an increasingly important part of the strategic leader's toolkit.

Cloud Infrastructure: Strategic Access to AI-Scale Compute

While algorithms and models receive most of the attention, AI's practical success depends fundamentally on the availability of scalable computational infrastructure. Machine learning, and deep learning in particular, demands significant processing power, high-throughput data ingestion, and robust deployment environments. Cloud computing has become the dominant operating model for AI development and deployment precisely because it provides elastic access to the specialized hardware and managed services that AI workloads require, without the capital expenditure and operational complexity of owning physical infrastructure.

Cloud platforms provided by Amazon Web Services (AWS), Microsoft Azure, and Google Cloud Platform allow enterprises to scale compute, storage, and networking resources elastically in response to project demands. AI training runs that would take weeks on local machines can be completed in hours using distributed GPU clusters; once training is complete, resources can be released instantly, avoiding idle capital expenditure. For inference workloads, cloud platforms provide managed serving environments with autoscaling, load balancing, monitoring, and versioning capabilities collectively known as MLOps or the operational discipline that ensures models remain consistent and reliable in production.

In addition to raw compute, cloud platforms provide higher-level managed AI services, including pre-trained foundation models, automated machine learning tools, vector databases, feature stores, and end-to-end development pipelines. These services dramatically reduce barriers to adoption by allowing organizations to leverage advanced AI capabilities without building them from scratch. A financial services firm can access state-of-the-art document processing through an API rather than training a custom model; a retailer can deploy a recommendation engine using a managed service rather than maintaining a bespoke ML platform.

However, managed services also require careful governance to avoid vendor lock-in, cost overruns, and uncontrolled model proliferation. The build-versus-buy-versus-managed-service decision is *one of the most consequential strategic choices* in AI infrastructure planning and should be made systematically for each infrastructure layer using a consistent decision framework. Key criteria include: competitive differentiation (does proprietary capability in this area create strategic advantage?), time-to-value requirements (is speed of deployment more important than customization?), total cost of ownership over a three-to-five-year horizon, and data governance constraints that may limit the use of third-party services for sensitive workloads. Organizations that apply this framework layer by layer, rather than defaulting to either full build or full managed services, will develop AI infrastructure portfolios that are both strategically differentiated and economically sustainable.

Few production AI environments operate exclusively on a single cloud platform. Hybrid architectures (e.g., combining public cloud, private cloud, and on-premises compute) are increasingly common, driven by data sovereignty requirements, latency constraints, cost optimization, and the desire to avoid single-provider dependency. Edge computing extends this architecture further, enabling AI inference to execute on IoT devices, mobile hardware, and on-premises systems where connectivity, latency, or regulatory requirements make cloud-only deployment impractical. For strategic planners, the key implication is that AI infrastructure is not a single technology choice but an architectural portfolio decision that must be revisited as workloads, regulations, and provider capabilities evolve.

FinOps for AI: Managing the Economics of Intelligent Infrastructure

Cloud platforms provide flexibility and scale, but they also introduce a cost structure that is fundamentally different from traditional IT infrastructure and that requires a distinct management discipline. AI workloads, characterized by variable compute intensity, unpredictable training run durations, and high-throughput inference demands, can generate cloud costs that escalate rapidly and unexpectedly if not actively managed. Business organizations that treat AI infrastructure costs as a technical detail managed by engineering teams, rather than as a strategic variable requiring financial governance, consistently experience cost overruns that undermine the business case for AI investment.

Financial Operations (FinOps) for AI is the practice of applying financial accountability and discipline to cloud and compute expenditure across the AI lifecycle. FinOps encompasses cost visibility or understanding what AI workloads cost at the model, team, and initiative level; cost optimization or identifying and implementing technical and organizational measures to reduce expenditure without sacrificing performance; and cost governance or establishing policies, accountability structures, and approval processes that prevent uncontrolled spending. For AI programs at scale, FinOps is not optional, but rather a prerequisite for financial sustainability.

AI infrastructure costs arise at every stage of the lifecycle and must be managed differently at each stage. During experimentation and development, the primary cost drivers are GPU compute for training runs and the storage of large training datasets. Costs at this stage are highly variable and difficult to predict, making budget controls and resource quotas important governance mechanisms. During production deployment, inference costs become the dominant ongoing expenditure, particularly for high-volume applications such as recommendation systems, NLP APIs, or real-time fraud detection that must serve millions of requests per day.

Technical optimization strategies that reduce inference cost without materially compromising model performance include model quantization (reducing numerical precision of model weights), pruning (removing low-importance model parameters), knowledge distillation (training smaller models to replicate the behavior of larger ones), and caching (storing frequently requested inference results rather than recomputing them). At the infrastructure level, organizations can reduce costs by using spot or preemptible compute instances for training workloads, right-sizing inference deployments to match actual traffic patterns and selecting hardware architectures calibrated to the latency and throughput requirements of specific applications rather than defaulting to the most powerful available option.

Effective FinOps for AI requires organizational structures that create accountability for AI infrastructure expenditure at the business unit, team, and initiative level. Chargeback and showback mechanisms, which allocate cloud costs to the organizational units that incur them, create the financial incentives for teams to optimize their AI resource usage rather than treating cloud compute as a shared resource with no individual accountability. Tagging policies that require every cloud resource to be labeled with the business initiative, team, and environment it supports provide the cost visibility that financial governance requires.

AI infrastructure investment governance should also include formal stage-gate reviews that evaluate the cost trajectory of AI initiatives at defined milestones and assess whether projected business value justifies continued expenditure. Initiatives that are consuming significant compute resources without demonstrating progress toward measurable business outcomes should be subject to explicit escalation and decision processes rather than continuing to consume budget by default. For executives, the key insight is that treating AI infrastructure investment with the same financial discipline applied to capital expenditure, i.e., with explicit business cases, return-on-investment tracking, and regular portfolio reviews, is essential for sustaining the organizational commitment to AI investment over the multi-year timelines that meaningful digital transformation requires.

Data Architecture at Scale: Centralized Platforms vs. Data Mesh

One of the most consequential and most actively debated decisions in enterprise AI infrastructure is how to organize data ownership, management, and access at scale. The choice between centralized data platform architectures and decentralized data mesh approaches reflects a fundamental tension between the efficiency of centralization and the domain expertise and organizational agility of decentralization. Getting this architectural choice right has *profound* implications for the speed of AI development, the quality of data available to models, and the organization's ability to maintain governance as the number and complexity of AI applications grows.

The centralized data platform model, exemplified by enterprise data warehouses, cloud data lakehouses, and unified analytical platforms, concentrates data storage, processing, and governance in a shared infrastructure managed by a central data engineering team. This approach offers significant advantages for governance, consistency, and the prevention of data silos: a single team is responsible for data quality standards, access controls, and infrastructure reliability, and all data consumers access the same authoritative data sources through a controlled interface.

The limitations of centralized models become apparent at scale. As data volumes grow and the number of data producers and consumers increases, central data engineering teams become bottlenecks: domain teams wait weeks for their data ingestion and transformation requests to be implemented by a team with limited bandwidth and limited domain knowledge. The central team's unfamiliarity with the specific characteristics and quality requirements of data from diverse business domains leads to quality issues that are difficult to diagnose and resolve. For AI applications that require deep integration between domain knowledge and data quality, centralized models can impede the rapid iteration that effective AI development demands.

Data *mesh*, a concept introduced by Zhamak Dehghani, proposes a fundamentally different organizational model for enterprise data management. Rather than centralizing data ownership in a platform team, data mesh distributes ownership to the domain teams such marketing, supply chain, finance, HR and those that are closest to the data and best positioned to understand its characteristics, quality requirements, and appropriate uses. Each domain team is accountable for its data as a product: maintaining quality, documenting characteristics, providing stable interfaces for consumers, and governing access in accordance with enterprise policies.

The data mesh model is supported by four organizing principles: *domain ownership*, which assigns data accountability to domain teams rather than a central platform; *data as a product*, which requires domain teams to treat their data outputs with the same quality and reliability standards applied to software products; *self-serve data infrastructure*, which provides shared platform capabilities that domain teams can use to publish and consume data without requiring custom engineering from a central team; and *federated computational governance*, which applies enterprise-level data policies through automated tooling rather than manual enforcement.

Neither centralized nor data mesh architectures are universally superior; the appropriate choice depends on organizational scale, domain complexity, data governance maturity, and the nature of AI use cases being pursued. Smaller organizations with limited data engineering capacity, highly standardized data environments, or AI use cases that require deep cross-functional data integration typically benefit from centralized approaches that concentrate scarce expertise and prevent fragmentation. Large, complex organizations with diverse business domains, high data production velocity, and AI programs that require deep domain-specific data quality are better candidates for data mesh approaches, provided they have the organizational maturity to implement distributed accountability effectively.

Many organizations find that a hybrid model, such as combining central platform infrastructure and governance with domain-level data ownership, provides the most practical balance. A central platform team provides the shared infrastructure (storage, compute, governance tooling, and access management) that all domains use, while domain teams own their data products and are accountable for their quality and reliability. For business and technology leaders, the strategic implication is clear: the data architecture decision is not purely technical. Instead, it is a question of organizational design, accountability structure, and management culture that requires executive ownership and explicit governance to resolve.

Data Catalogs and Metadata Management: Making Data Discoverable

One of the most pervasive and underappreciated barriers to effective AI development in large organizations is not data quality or data infrastructure but rather data discoverability. Data engineers and data scientists working on AI applications frequently spend a substantial fraction of their time not building models, but searching for the data they need: determining whether a relevant dataset exists, where it is stored, who owns it, what it contains, how current it is, and whether it is of sufficient quality for their intended use. In organizations with dozens of data systems, hundreds of datasets, and thousands of potential users, this discovery problem is severe enough to materially slow AI development velocity and increase the cost of AI programs.

Data *catalogs* address this problem by providing a centralized, searchable inventory of an organization's data assets, enriched with the metadata that data consumers need to evaluate fitness for purpose. A well-maintained data catalog enables a data scientist beginning a new AI project to discover in minutes what would otherwise require days of internal networking: what datasets exist, their schema and contents, their provenance and update frequency, their documented quality characteristics, their access policies, and who to contact with questions.

Modern enterprise data catalog platforms, including tools such as Alation, Collibra, DataHub, and Atlan, provide several interconnected capabilities. Automated metadata ingestion connects to data sources across the enterprise and extracts structural metadata such as table schemas, column names, data types, row counts, and update timestamps all without requiring manual documentation effort. Business glossary and semantic tagging allow organizations to annotate datasets with business-meaningful descriptions, ownership information, and classification labels (such as sensitivity classifications for personal or regulated data) that make catalog entries interpretable to non-technical users.

Data *lineage tracking* or the ability to trace the origin of a dataset, the transformations applied to it, and the downstream systems and models that consume it represents a particularly critical capability for AI governance. When a production AI model produces unexpected outputs, lineage tracking enables rapid diagnosis: which training data contributed to the model, how was it processed, and did any upstream data quality issues propagate into the training set? Without lineage tracking, this diagnosis requires time-consuming manual investigation that delays both issue resolution and regulatory response.

For business and technology leaders, the case for investing in data catalog infrastructure is both operational and strategic. Operationally, data catalogs reduce the time data teams spend on data discovery, accelerate the onboarding of new team members, and reduce the risk of data quality issues propagating into AI models undetected. Strategically, well-maintained catalogs create institutional knowledge about data assets that persists through staff turnover, enables better data governance by making it easier to apply access controls and sensitivity classifications consistently, and provides the audit trail that regulators and internal auditors increasingly require for AI systems deployed in high-stakes domains.

Data Contracts: Engineering Accountability into Data Pipelines

Among the most common and consequential causes of AI model failures in production is a problem that is rarely visible during model development: upstream data changes that are not communicated to downstream consumers. A data engineering team modifies the schema of a source table, renames a field, changes the granularity of an aggregation, or alters the logic of a transformation and then an AI model trained on the previous data format begins producing degraded or nonsensical outputs. By the time the degradation is detected through model monitoring, the impact may already have reached end users, customers, or business decisions.

Data contracts are a mechanism for preventing this class of failure by formalizing the interface between data producers, essentially the teams or systems that generate data, and data consumers or the teams and systems that use it. A data contract specifies, in machine-enforceable terms, the schema, data types, null constraints, value ranges, update frequency, and quality guarantees that a data producer commits to maintaining for a defined period. When a producer intends to make a change that would violate an existing contract, the contracting framework requires explicit negotiation with affected consumers before the change is deployed, preventing silent breaking changes.

Data contracts are typically expressed as code or schema definitions, validation rules, and quality assertions embedded in the data pipeline infrastructure and then enforced through automated testing in CI/CD pipelines that prevent non-compliant data from propagating downstream. Platforms such as **dbt** (data build tool) have popularized the use of data quality tests as first-class artifacts of the data transformation process, establishing a foundation for more formal contractual frameworks. Dedicated data contract platforms, including Soda, Great Expectations, and emerging purpose-built contract management tools, extend this foundation with explicit producer-consumer agreement tracking, version management, and breach notification systems.

For business and technology leaders, the strategic importance of data contracts extends beyond their technical function. Data contracts operationalize the concept of data as a product: just as software engineering teams commit to stable APIs for application consumers, data engineering teams commit to stable data interfaces for AI and analytics consumers. This commitment requires organizational structures that assign clear ownership of data products, governance processes that require impact assessment before data changes are deployed, and cultural norms that treat data consumers as stakeholders in data design decisions rather than passive recipients of whatever data is available. Organizations that invest in data contracts as an organizational practice and not merely a technical tool will substantially reduce the operational cost and risk of maintaining AI systems in production.

Real-Time and Streaming Data Infrastructure

Many of the highest-value AI applications require decisions to be made in real time such as fraud detection "at the moment" of a transaction, personalized recommendations at the instant of a user interaction, predictive maintenance alerts before equipment failures propagate. These applications cannot be adequately served by traditional batch-oriented data architectures in which data is collected, processed, and made available on hourly or daily cycles. They require streaming-first infrastructure: systems designed to ingest, process, and serve data continuously, with latency measured in milliseconds to seconds rather than minutes to hours.

The shift to streaming-first architectures represents one of the *most significant infrastructure transitions* in the modern data stack, and one that many organizations are still navigating. Batch systems are simpler to build, easier to debug, and sufficient for a wide range of analytics and AI applications that do not require real-time decision-making. However, as the proportion of AI use cases requiring real-time or near-real-time data grows, organizations must develop competence in streaming data infrastructure or accept a structural constraint on the AI applications they can support.

The streaming infrastructure stack comprises several interconnected components, each serving a distinct function in the real-time data pipeline. Event streaming platforms (of which Apache Kafka is the dominant open-source example, with managed equivalents available through AWS Kinesis, Azure Event Hubs, and Google Pub/Sub) provide durable, high-throughput message buses that decouple data producers from consumers and enable multiple consumers to independently process the same event stream. These platforms guarantee that events are not lost even when downstream consumers are temporarily unavailable, and they provide the replay capability that is essential for retraining AI models on historical event sequences.

Stream processing engines, including Apache Flink, Apache Spark Streaming, and managed equivalents, now enable real-time transformation, aggregation, and enrichment of data as it flows through the streaming pipeline. These engines apply the transformation logic that in batch architectures would be performed in overnight ETL jobs, but in real time and with the low latency required for AI inference. For AI-specific applications, stream processing engines compute the features that production models consume at inference time such as aggregating behavioral signals over sliding time windows, joining event streams with reference data, and publishing computed features to the online feature store that serves the model.

A challenge specific to AI systems in streaming environments is the consistency of features between training and inference. Models are typically trained on historical data processed in batch, but they are served in production using features computed in real time from streaming data. If the feature computation logic differs between the batch and streaming paths (perhaps due to implementation differences, timing discrepancies, or data source inconsistencies) then the model's production performance will be systematically degraded relative to its training-time performance, a phenomenon known as *training-serving skew.*

Online feature stores, meaning shared infrastructure that provides a consistent API for writing features computed in streaming pipelines and reading them at inference time, can address this challenge by maintaining a synchronized view of the feature values that models should receive. Features computed by stream processing engines are written to the online feature store in real time; model serving infrastructure reads from the same store at inference time, guaranteeing that the features the model receives match those on which it was trained. For business organizations building AI applications at scale, the ability to maintain training-serving consistency across batch and streaming environments is a critical infrastructure capability that requires explicit architectural planning.

Regulatory Constraints and Data Residency: Infrastructure Implications

AI infrastructure decisions do not take place in a regulatory vacuum. The regulatory environment governing data collection, storage, processing, and transfer is extensive and continues to evolve rapidly across jurisdictions, and it places material constraints on the architectural choices available to organizations operating across geographies. Business and technology leaders who treat data residency and regulatory compliance as technical details to be resolved by their legal and engineering teams, rather than as strategic inputs to infrastructure design, frequently discover that regulatory constraints have invalidated infrastructure investments made without adequate compliance analysis.

The most significant categories of regulatory constraint for AI infrastructure planning are data residency requirements, data transfer restrictions, sector-specific data handling obligations, and AI-specific governance regulations. Each category imposes different architectural constraints and requires different organizational responses.

Data *residency regulations* require that certain categories of data (typically personal data of residents of the regulating jurisdiction, health records, financial data, or government-related data) be stored and processed within defined geographic boundaries. The European Union's General Data

Protection Regulation (GDPR) imposes stringent conditions on the transfer of personal data outside the EU, requiring that recipient jurisdictions provide adequate data protection or that specific transfer mechanisms such as Standard Contractual Clauses be employed. China's Data Security Law and Personal Information Protection Law impose similar requirements for data collected within China, with additional requirements for data related to national security or critical information infrastructure.

For business organizations operating AI systems across multiple jurisdictions, these requirements have concrete architectural implications. Training data for AI models must be stored in compliant locations and cannot be freely centralized in a single cloud region without assessing whether cross-border transfer is permissible. Model training itself may need to be conducted in the jurisdiction where training data is stored, or federated learning techniques may be required to train models without centralizing data. Inference infrastructure, i.e., the systems that serve model predictions to end users, may need to be deployed regionally to ensure that user data does not leave jurisdictional boundaries. Cloud provider selection must account for the availability of compliant data center regions, and contracts with cloud providers must address data processing terms required by relevant regulations.

Beyond general data protection regulations, sector-specific frameworks impose additional requirements that shape AI infrastructure design in specific industries. In healthcare, HIPAA in the United States requires that protected health information be handled with specific administrative, technical, and physical safeguards that constrain cloud infrastructure choices, access control architectures, and audit logging requirements. In financial services, regulations including PCI-DSS for payment card data, SOX for financial reporting systems, and sector-specific regulations from bodies such as the SEC, FCA, and ECB impose data integrity, audit trail, and system resilience requirements that directly affect how AI systems are designed, deployed, and monitored. In the European Union, the AI Act imposes risk-based governance requirements on AI systems deployed in high-risk categories including employment, credit, healthcare, and critical infrastructure, with documentation, transparency, and human oversight obligations that affect how AI systems are architected and operated.

For business and technology leaders, the practical implication is that regulatory compliance must be integrated into AI infrastructure design from the outset, not retrofitted after deployment. Infrastructure architects should work with legal and compliance teams at the *design stage* of AI initiatives to identify applicable regulatory requirements, assess their architectural implications, and incorporate compliant patterns into the infrastructure design. Business organizations that build AI infrastructure with regulatory requirements in mind from the start avoid the costly and disruptive compliance remediation that organizations face when regulatory requirements are discovered after deployment.

Security in AI systems extends beyond conventional cybersecurity to encompass threats specific to AI architectures: data poisoning (inserting malicious examples into training data to manipulate model behavior), model extraction (querying a model repeatedly to reconstruct its architecture or training data), prompt injection (crafting inputs that override the intended instructions of language model systems), and adversarial examples (inputs crafted to cause systematic model misclassification). Understanding these threats is a prerequisite for designing AI systems that are robust in adversarial deployment environments.

Red-teaming or the practice of systematically attempting to cause AI systems to fail by simulating adversarial inputs and attack strategies has emerged as an essential operational practice for AI systems deployed in high-stakes or public-facing contexts. Red-teaming exercises identify failure modes that are not apparent through standard evaluation benchmarks, including jailbreaking vulnerabilities in language model systems, demographic bias patterns in computer vision systems, and distributional shift vulnerabilities in predictive models. Leading AI business organizations conduct red-teaming exercises before deploying AI systems in production and establish ongoing red-teaming programs that continue to probe deployed systems as adversarial techniques evolve. For executives, the key implication is that adversarial robustness is not a one-time certification but a continuous operational discipline that requires dedicated resources and organizational accountability, analogous to continuous penetration testing in conventional cybersecurity programs.

Organizational Roles in Data Governance: Making Accountability Real

Data governance frameworks frequently fail not because they are poorly designed in principle but because the organizational structures required to implement them are absent. Governance policies that specify data quality standards, access control requirements, and lineage documentation obligations are only as effective as the organizational roles responsible for implementing and enforcing them. Business organizations that publish data governance frameworks without defining the accountable roles and embedding those roles in organizational structures with real authority and resources find that governance remains aspirational rather than operational.

Three organizational roles are foundational to effective data governance for AI: the *data owner*, the *data steward*, and the *data product manager*. Each role occupies a distinct position in the governance structure and performs functions that the other roles cannot. Understanding these roles, especially their responsibilities, their authority, and their relationships to each other and to the AI development teams that depend on governed data, is essential for business and technology leaders designing or assessing their organization's AI readiness.

The data owner is a senior business leader, typically at the director or vice-president level, who holds executive accountability for a specific domain's data assets. Data ownership is a business accountability, not a technical one: the data owner is responsible for ensuring that data from their domain is fit for the purposes for which it is used across the organization, that access policies reflect appropriate risk management, and that data governance obligations are met. Data owners have the authority to make binding decisions about data access, data sharing, and data quality standards for their domain, and they are accountable to senior leadership for the quality and governance of their data assets.

Data stewards are practitioners, e.g., data engineers, business analysts, or subject matter experts who implement the governance decisions made by data owners and manage the day-to-day quality and documentation of specific datasets. While data owners set governance policy, data stewards execute it: maintaining data dictionaries, managing data quality monitoring, documenting data lineage, reviewing and approving access requests, and serving as the primary point of contact for consumers who have questions about specific datasets. In business organizations adopting data mesh architectures, data stewards are often embedded within domain teams, providing governance expertise close to the data and the business context in which it is generated.

The data product manager is a newer role that has emerged in business organizations adopting data mesh or product-oriented data management approaches. Modeled on the software product manager role, the data product manager is responsible for the strategy, roadmap, quality, and adoption of a specific data product, a curated, well-documented, and reliably maintained data asset that serves multiple consumers. Data product managers work with data owners and data stewards to define what data products should exist, what quality standards they should meet, and how they should evolve in response to consumer needs. They measure the success of data products by their adoption, their reliability, and the business outcomes they enable all while applying the product management discipline to data assets rather than software features.

For business and technology leaders, the key strategic insight is that effective data governance for AI requires organizational investment in these roles and not merely the publication of governance policies or the purchase of governance tooling. Organizations that appoint data owners with genuine authority, embed data stewards with adequate time and expertise, and develop data product manager capabilities create the organizational infrastructure that translates governance frameworks into operational reality. Those that do not will find that governance remains a nominal commitment rather than an operational capability, with predictable consequences for AI model quality, regulatory compliance, and the trust of the stakeholders who depend on AI-informed decisions.

Conclusion: Technology Literacy as a Leadership Competency

The technologies examined in this chapter, namely machine learning and its three principal paradigms, deep learning, natural language processing, computer vision, and the hardware and infrastructure architectures that enable them together constitute the foundational vocabulary of AI-fluent leadership. But as this chapter has demonstrated, effective AI leadership also requires understanding the organizational and economic dimensions of AI infrastructure: the data architecture choices that determine how data is owned and governed at scale, the financial discipline required to manage AI compute costs responsibly, the regulatory constraints that shape where data can be stored and how AI systems must be designed, the data contracts and catalog infrastructure that make data assets discoverable and reliable, the streaming infrastructure that enables real-time AI applications, and the organizational roles that translate governance frameworks from policy into practice.

The most consequential mistakes in enterprise AI programs are rarely technical failures. They are *failures of strategic alignment* between technology choices and business objectives, failures of financial governance that allow AI costs to grow faster than AI value, failures of data governance that allow model quality to degrade undetected, and failures of organizational design that leave governance frameworks without the accountable roles required to implement them. Business and technology leaders who develop literacy across both the technical and business organizational dimensions of AI infrastructure are better positioned to avoid these mistakes and to build the resilient, scalable, and trustworthy AI capabilities that true digital transformation and competitive advantage require.

CHAPTER 3

The Valley of Despair: Why Enterprise AI Initiatives Fail

ARTIFICIAL INTELLIGENCE DELIVERS ITS GREATEST value when it is applied to real business problems within a structured project life cycle. Organizations that succeed with AI follow a disciplined, iterative process, one that emphasizes business alignment, data quality, technical rigor, and operational reliability. Equally important is the development of an AI-ready organization that brings together the right roles, cultural norms, and operating structures. The following section illustrates how these elements come together through practical, business-oriented use cases across core functional domains.

The AI Project Life Cycle in Practice

Modern AI development differs significantly from traditional software engineering. Whereas conventional systems are designed to execute deterministic logic, AI systems learn probabilistic patterns from data. As a result, the AI life cycle is cyclical and experimental, requiring continuous iteration, feedback, and monitoring. The real-world use cases below demonstrate how this process unfolds in business environments.

The cycle begins with precise business *scoping*. Consider a global manufacturing firm attempting to reduce unplanned downtime on a high-speed milling machine. Rather than pursuing a vague objective such as "improving machine reliability," business leaders and AI engineers narrow the problem to a concrete prediction task: estimating the probability of machine failure within the next 72 hours. This formulation is powerful because it aligns the technical goal with measurable financial outcomes or downtime hours avoided, maintenance costs saved, and production throughput improved. When scoping is done correctly, every subsequent step of the AI life cycle is grounded in business value.

Once the problem is clearly defined, organizations confront the most resource-intensive stage of the project: data acquisition and preparation. In the manufacturing example, data engineers and data scientists work together to combine sensor readings, maintenance histories, operator logs, and environmental measurements. A significant portion of the project timeline is consumed by cleaning corrupted timestamps, dealing with missing entries, normalizing sensor readings, engineering meaningful features, and labeling historical failures. Teams quickly learn that the performance of a predictive maintenance model depends far more on subtle features extracted from vibration patterns or temperature trends than on the choice of modeling algorithm. As a general rule good data, not sophisticated algorithms, determines success.

With data prepared, the project transitions into the modeling and evaluation phase. Data scientists explore multiple supervised learning techniques, assessing gradient boosting models, deep neural networks, and ensemble approaches. However, the selection of the "best" model is not driven by accuracy alone. Instead, stakeholders use a cost framework that weighs the financial impact of prediction errors. A model that slightly underperforms technically may be preferred if it minimizes false negatives that could lead to catastrophic equipment failure. In this way, business metrics and not technical metrics guide decision-making.

The final stage of the life cycle involves deployment and monitoring, commonly managed through MLOps practices. Returning to the manufacturing example, AI engineers integrate the chosen model with the factory's monitoring dashboards and alert systems. The model ingests real-time sensor data and automatically generates maintenance tickets when the probability of failure exceeds a threshold. Continuous monitoring is essential because real-world conditions constantly change, for example machines age, sensors drift, and production patterns shift. Without mechanisms for retraining and recalibration, model performance can degrade over time. In AI, deployment is not the end of the project but the beginning of ongoing operational stewardship.

Building the AI-Ready Organization

Even the most sophisticated model will fail if the organization around it is unprepared. Effective AI implementation relies on cross-functional teams, supportive culture, and clear operating structures. Successful projects begin with three essential roles: 1) *Data scientists* design and train models, applying expertise in statistics and programming to extract patterns from data; 2) *AI engineers* build the pipelines, APIs, and infrastructure that allow models to run reliably at scale, ensuring they can be deployed into production systems; and 3) *AI translators*, sometimes known as product owners or business liaisons, connect technical work to organizational goals. They define the business problem, interpret model outputs for decision-makers, coordinate with data owners, and guide change management. The combination of these three roles creates a balanced team capable of executing AI initiatives from conception to production.

Technical expertise alone, however, is not enough. Organizations must cultivate a culture that embraces experimentation and uncertainty. Because AI predictions are probabilistic, perfect accuracy is impossible. Teams must be willing to pilot models, evaluate their performance, and iterate based on results. Leadership plays a key role by rewarding evidence-based decision-making and encouraging controlled experimentation even when outcomes are uncertain. A data-centric culture also requires treating data as a strategic asset, rather than as a byproduct of operations. Data governance, quality control, and accessibility become shared responsibilities across departments, not simply confined to IT.

Organizational structure also shapes the success of AI initiatives. Many enterprises explore different models for arranging their AI talent. Some adopt a centralized structure, where a single corporate data science team manages all initiatives. While this approach promotes standardization, it may limit speed and domain-specific learning. Others adopt a decentralized structure, embedding data scientists directly within business units such as marketing, operations, or finance. This structure accelerates domain expertise but carries the risk of inconsistent practices and siloed work. The most effective model for large organizations is often a federated, or hub-and-spoke, approach. A central AI *Center of Excellence* provides governance, infrastructure, and advanced expertise, while smaller embedded teams develop localized solutions informed by deep domain knowledge. This hybrid structure supports both innovation and consistency.

Key AI Applications in Functional Domains

The integration of AI into the modern enterprise represents a shift from static, rule-based processes to dynamic, data-driven strategies. Once an organization has established a functional AI life cycle, i.e., moving from data collection to model deployment, it can begin to harvest transformative value across various departments. This section describes how machine learning and related technologies enhance decision-making and operational efficiency within specific business functions.

In the domains of marketing and sales, AI serves as the primary engine for hyper-personalization, moving away from traditional, one-size-fits-all campaigns. Streaming services provide a premier example of this shift by employing machine learning to analyze granular data points, including individual viewing patterns, device preferences, and specific time-of-day habits. By processing this interaction history through *recommendation engines*, these companies can tailor content suggestions to the unique tastes of each user, which directly correlates to increased customer engagement and long-term retention.

Moreover, in a B2B context, AI optimizes the sales funnel through *lead-scoring models*. These models prioritize sales outreach by analyzing behavioral signals such as website visits, email interaction rates, and historical touchpoints. By assigning a "likelihood-to-convert" score to every prospect, AI allows sales teams to focus their energy on the most promising leads coming through the pipeline. This targeted approach ensures that organizational resources are used efficiently, ultimately closing more deals without necessarily increasing headcount.

Operational functions benefit immensely from the predictive capabilities of AI, which allow firms to anticipate market needs and technical failures. Retailers, for example, utilize advanced *demand-forecasting models* that synthesize historical sales with complex external variables like seasonality, weather patterns, and local events. These models are significantly more accurate than traditional statistical methods, enabling companies to maintain optimal inventory levels, reduce product spoilage, and ensure that items are available on the shelf when the customer needs them.

Beyond forecasting, AI is critical for *predictive maintenance*, a strategy that utilizes sensor data like capturing temperature, vibration, and pressure to monitor the health of high-value assets like aircraft engines. Machine learning models can detect subtle signs of component degradation long before a physical failure occurs, allowing maintenance to be scheduled proactively. This ultimately reduces operational disruptions, lowers long-term repair costs, and enhances overall safety. Furthermore, reinforcement learning is being applied to logistics to optimize route selection for delivery fleets, which continuously improves delivery times and reduces fuel consumption.

In finance and HR, AI acts as a sophisticated tool for risk mitigation and strategic planning. Financial institutions utilize *anomaly-detection algorithms* to analyze millions of transactions in real time to identify the subtle patterns of fraudulent activity. Because these models can adapt to new tactics faster than rigid, rule-based systems, they are essential for reducing financial losses in an evolving threat landscape.

Within Human Resources, large-scale enterprises apply Natural Language Processing to streamline the recruitment process. Natural Language Processing allows systems to analyze resumes, match candidate qualifications to specific job descriptions, and identify the most promising talent with speed and a higher degree of fairness. Additionally, AI aids in workforce stability by predicting which employees may be at risk of leaving the company, providing HR leaders with the insights needed to address retention challenges before they impact the organization.

Human-in-the-Loop and Economic Realities: Why AI Projects Fail Where It Matters Most

There is a particular category of AI project failure that is especially instructive because it occurs not in the early experimental stages, where failure is expected and relatively inexpensive, but at the precise moment when a project appears to have succeeded. The model has been built, validated, and deployed. The technical team has demonstrated performance metrics that meet or exceed the targets established at project inception. The pilot results were compelling enough to secure funding for enterprise rollout. And then, somewhere in the transition from controlled pilot to operational reality, the project quietly stops delivering the value that justified its existence; and not because the technology failed, but because the organizational and economic architecture surrounding the technology was never designed to support the kind of human engagement that separates a functioning AI system from a genuinely valuable one.

This failure pattern (so common that it has become a recognizable archetype in enterprise AI program management) is rooted in two interconnected realities that AI project planning systematically underestimates: the genuine complexity and cost of meaningful human-in-the-loop design, and the economic dynamics that make sustained human oversight of AI systems far more difficult to maintain in practice than the project business case assumed it would be. Understanding these realities not as edge cases or implementation details but as fundamental structural challenges that must be designed for from the beginning is among the most important lessons that the accumulating record of enterprise AI project outcomes can teach.

The concept of *human-in-the-loop*, at its most straightforward, refers to the design principle that consequential AI system outputs should be reviewed, validated, or approved by a human decision-maker before they are acted upon. In practice, however, the term encompasses a spectrum of human engagement models that differ dramatically in their implications for system design, organizational workflow, regulatory compliance, and economic viability. At one end of the spectrum is human-in-the-loop in its strictest sense, meaning a design in which every AI output is reviewed by a human before action is taken; the human has genuine authority to override the AI recommendation and the system is explicitly designed to support that review with sufficient context, explanation, and time for meaningful human judgment to operate.

At the other end is what practitioners sometimes call human-on-the-loop there is a design in which AI systems operate autonomously within defined parameters; humans monitor aggregate system behavior and retain the ability to intervene at the system level, but individual outputs are not reviewed before they generate consequences. Between these two poles lies a rich and consequential design space, and where an organization positions its AI systems within that space, and whether that positioning is deliberate or accidental, genuinely implemented or merely documented, is one of the most important determinants of both the value the system delivers and the risk it generates. The failure mode that recurs most consistently in enterprise AI deployments is not that organizations make the wrong choice about where on this spectrum their systems should sit. But rather that they make a choice in the design phase that sounds responsible and is recorded in the governance documentation, and then they discover during deployment that the operational and economic conditions required to implement that choice do not exist and were never created.

The gap between designed and actual human-in-the-loop engagement is produced by a set of cognitive, behavioral, and organizational dynamics that are well documented in the research literature and entirely predictable in advance, yet for some reason are persistently absent from AI project risk registers. The first and most thoroughly studied is *automation bias* or the human tendency to over-rely on automated system recommendations, accepting AI outputs without the critical scrutiny that the human-in-the-loop design assumes will be applied, particularly when the AI system has established a track record of apparent accuracy that builds misplaced confidence in its reliability. Automation bias is not a character flaw of the individuals who exhibit it, but rather a predictable cognitive response to the experience of reviewing large volumes of AI recommendations under time pressure, where the effort required to genuinely evaluate each recommendation against independent judgment is unsustainable, and where the feedback mechanisms that would reveal the AI's errors are delayed, indirect, or absent.

A radiologist reviewing AI-flagged medical images under the production volume pressures of a busy hospital department, a loan officer processing AI-generated credit recommendations against the throughput targets of a retail lending operation, a content moderator evaluating AI-classified social media posts at the scale that platform economics demand.... each of these human-in-the-loop roles is designed with the assumption that the human reviewer will exercise genuine independent judgment on every case. Each is structured in ways that make genuine independent judgment practically impossible to sustain, and that predictably produce the automation bias that converts a human-in-the-loop design into a de facto automated system with human rubber-stamping as a compliance theater layer. When that automated system makes a consequential error, e.g., misdiagnosing a malignancy, declining a creditworthy borrower, allowing harmful content to remain on the platform, the human-in-the-loop design that was supposed to prevent exactly this outcome provides neither the safety it promised nor the regulatory protection it was intended to establish.

The economic realities that compound these cognitive dynamics create a structural tension at the heart of human-in-the-loop AI design that most project business cases never honestly confront. The economic justification for AI deployment in most enterprise contexts rests on some combination of cost reduction through automation, revenue enhancement through improved decision quality, and risk reduction through more consistent application of decision criteria. All three of these value sources are maximized (in the short term) by minimizing the human involvement in AI-assisted decision processes: fewer reviewers mean lower labor cost, faster throughput means higher revenue processing capacity, and more consistent AI application of decision rules means less variance from human judgment idiosyncrasy.

The human-in-the-loop design requirement pulls in exactly the opposite direction: meaningful human oversight requires human time, human expertise, human organizational infrastructure, and human management attention, all of which have real costs that the project ROI model must absorb. This tension is manageable when it is acknowledged honestly in the project design phase and the business case is constructed around a realistic model of what human oversight will in-fact cost to implement correctly. However, it becomes a project-destroying dynamic when the business case is built around the automation value and the human oversight cost is treated as negligible (because it is only in theory, not in the financial model) and the organization subsequently discovers that implementing meaningful oversight requires investment that the approved budget does not contain and that the achieved ROI cannot support. At this point, the organization faces a choice between implementing oversight at a level that is genuine but more expensive than the business case assumed, implementing oversight at a level that is affordable but inadequate, or quietly abandoning the oversight requirement and accepting the risk and regulatory exposure that situation entails. In the majority of documented cases, organizations choose the second option, i.e., implementing a form of human oversight that satisfies the letter of the governance requirement while delivering none of its substance, and the project delivers returns that are lower than projected, risks that are higher than acknowledged, and outcomes that are worse for the people affected by AI decisions than the human-in-the-loop design was intended to ensure.

The economic realities of AI project failure extend beyond the human oversight cost problem to encompass a broader set of financial dynamics that the standard AI project business case architecture is structurally ill-equipped to capture. The first dynamic is the hidden cost of failure at scale, meaning the recognition that AI systems, unlike human decision-makers, fail systematically rather than randomly, meaning that when an AI model produces an incorrect output it tends to do so consistently across all cases sharing the characteristics that triggered the error, potentially generating large volumes of harmful decisions before the failure mode is detected. The cost of remediating systematic AI failures (e.g., identifying affected cases, reversing incorrect decisions, compensating affected individuals, managing regulatory response, and repairing reputational damage) is rarely modeled in AI project business cases because the probability of a systematic failure of the deployed model is treated as too low to warrant inclusion in the base case financial model.

The second dynamic is the cost of model degradation over time such as the well-documented phenomenon by which AI model performance decays as the real-world environment the model was trained to represent changes in ways that reduce the relevance of the training data. A fraud detection model trained on transaction patterns from 2022 will gradually lose accuracy as fraud methodologies evolve. A demand forecasting model trained on pre-pandemic consumer behavior will produce increasingly unreliable outputs as structural shifts in consumer patterns make historical data less predictive of future behavior. Maintaining model performance over time requires ongoing investment in monitoring, retraining, and validation that is distinct from the initial deployment investment and that must be sustained indefinitely for as long as the model remains in production. This fact represents an ongoing operational cost that is frequently either omitted from the project business case or estimated at levels that prove materially insufficient.

The third dynamic involves the economic reality defined as the opportunity cost of organizational attention or the recognition that AI projects, particularly during the deployment and scaling phases, consume management bandwidth, technical capacity, and organizational change energy that could be deployed against alternative priorities, and that this opportunity cost is real even when it does not appear as a line item in the project financial model.

Organizations that deploy multiple AI initiatives simultaneously without adequate governance of their collective demand on organizational capacity frequently find that each individual initiative is starved of the management attention required for successful adoption, producing a portfolio of mediocre outcomes that collectively deliver less value than a smaller number of well-resourced and well-managed initiatives would have achieved.

The path through these failure dynamics is not to abandon human-in-the-loop design principles or to accept that the economic realities of AI deployment make genuine human oversight impossible. The path is to build AI project business cases and governance frameworks that are honest about both the costs and the necessity of meaningful human engagement, and that design the operational conditions required for that engagement to be real rather than performative. This means staffing human review functions at levels consistent with genuine review rather than rubber-stamping throughput, designing feedback mechanisms that surface AI errors quickly enough to interrupt systematic failure before it generates large-scale harm, building model monitoring and retraining budgets into the operational cost structure from the beginning rather than treating them as future discretionary investments, and establishing governance processes that regularly assess whether the human-in-the-loop design documented at project inception is actually being implemented in operational practice.

Additionally, it also means being willing to reject AI deployment proposals whose business cases are only viable if human oversight costs are excluded, because an AI system deployed without adequate human oversight is not a less expensive version of a well-governed AI system. This is a categorically different system with a different risk profile, a different regulatory status, and a different set of consequences for the people whose lives it affects. The organizations that internalize this distinction early, and build their AI investment frameworks around it, will avoid the category of failure that is most damaging precisely because it arrives dressed as success.

CHAPTER 4

Developing an Enterprise AI Strategy

The Strategic Imperative of Artificial Intelligence

IN RECENT YEARS, ORGANIZATIONS ACROSS industries have recognized the transformative potential of artificial intelligence. Yet despite nearly universal enthusiasm, only a minority have produced AI strategies capable of generating sustained economic value. As the McKinsey Global Institute observed in 2025, most companies have an AI strategy, but very few have a strategy that will in-fact produce economic value. This observation underscores a crucial distinction: an AI strategy is not simply a technology roadmap drafted by the Chief Technology Officer, nor a collection of ambitious pilot projects scattered across business units. Rather, AI strategy is fundamentally a *business* strategy, one that is designed, owned, and governed by the highest levels of leadership and that deploys AI as a lever to achieve measurable financial and operational outcomes.

When developed effectively, an AI strategy becomes a central pillar of corporate transformation, aligning resources, priorities, and capabilities around durable value creation. A valid AI strategy disciplines an organization to make explicit choices: which problems are worth solving, which markets to pursue, which capabilities to build or acquire, and how to sequence investments to balance near-term impact with long-term transformation. Without this strategic framing, AI initiatives frequently devolve into experiments that generate technical insight but minimal economic return or what practitioners have come to call "pilot purgatory."

In this chapter we will consider a comprehensive framework for developing an enterprise AI strategy. The approach is organized around the key decisions, disciplines, and design principles that distinguish strategies capable of producing measurable impact from those that remain aspirational. The target audience for this material includes business and technology executives accountable for AI investment decisions, as well as graduate students studying the intersection of artificial intelligence, strategy, and organizational transformation.

Why Conventional Technology Planning Falls Short

The most common failure mode in enterprise AI initiatives is treating it as a technology program rather than a business strategy. When AI is positioned as a domain owned exclusively by the technology function, it often generates technically impressive outcomes that are organizationally disconnected from strategic priorities. Projects are evaluated on model accuracy rather than business impact. Governance is treated as a compliance obligation rather than a strategic enabler. Executive sponsorship is intermittent rather than sustained.

Valid and effective AI strategy requires a fundamentally different posture, one that begins with business imperatives such as the competitive pressures, customer demands, cost structures, and growth opportunities that define organizational success. Then the strategy works backward to identify where AI can serve as a decisive lever. This distinction, between technology-led and strategy-led AI, has measurable consequences. Organizations that embed AI into their corporate strategy, with C-suite accountability and board-level visibility, consistently outperform those that relegate AI to functional department or IT budgets.

Strategy serves as the indispensable bridge between technical potential and business imperatives. A well-constructed AI strategy performs several essential functions simultaneously. First, it forces organizational leaders to articulate which problems are genuinely worth solving at scale. Second, it establishes criteria for prioritizing investments across a landscape of competing opportunities. Third, it defines the organizational structures, talent models, and governance mechanisms needed to translate AI capability into sustained advantage. Moreover, it creates accountability systems that connect AI outputs to the financial and operational outcomes that matter to stakeholders.

Critically, AI strategy is not a one-time planning exercise. The rapid pace of model development, including the emergence of large language models, multimodal AI, and autonomous agent architectures, means that strategies must be adaptive, incorporating systematic mechanisms for scanning the technology landscape and fluidly reassessing priorities as capabilities evolve. The strategic planning cycle for AI should be *tighter* than traditional annual planning processes, with formal reassessment triggered by significant model capability shifts, competitive moves, or changes in the regulatory environment.

Research consistently demonstrates that strategic clarity in AI correlates with measurable competitive advantage. Companies that define a coherent

AI strategy, i.e., one that specifies target domains, capability development pathways, and value measurement frameworks, are significantly more likely to achieve at-scale deployment and report positive material financial impact. Conversely, organizations that pursue AI opportunistically, without strategic prioritization, tend to spread resources thinly across many initiatives, producing incremental improvements rather than true transformative outcomes.

The competitive stakes are particularly acute in industries where AI can redefine customer expectations, unit economics, or competitive barriers. In financial services, AI-driven underwriting and fraud detection can compress risk costs by double-digit percentages. In healthcare, diagnostic AI and predictive care management can simultaneously improve patient outcomes and reduce operational expenses. In retail, dynamic pricing and demand forecasting powered by machine learning can significantly improve inventory efficiency. Across these domains, organizations that develop early strategic mastery of AI will accumulate data assets, model expertise, and organizational capability that become increasingly difficult for competitors to replicate, thus creating durable, compounding advantage.

Identifying and Prioritizing High-Value Use Cases

Identifying high-value use cases is the first substantive step in building an AI strategy. Organizations frequently apply the principle of concentration, meaning seeking the small number of opportunities likely to deliver outsized returns while deliberately avoiding the temptation to pursue dozens of unfocused pilots. The Pareto principle applies with particular force in AI strategy: in most organizations, a small fraction of potential use cases will account for the vast majority of available economic value. The *strategic discipline* is in finding and funding those use cases while deprioritizing the rest.

High-value AI opportunities tend to share a consistent set of characteristics. They typically address processes that are data-rich, decision-intensive, and operationally significant. They involve decisions made at high frequency, where even modest improvements in decision quality compound into substantial aggregate impact. Their outcomes can be measured in clear business terms (e.g., revenue uplift, cost reduction, risk mitigation, or customer experience improvement) rather than proxy technical metrics. Moreover, they are sufficiently mature that viable modeling approaches and sufficient training data exist to support deployment within a reasonable timeframe.

Organizations benefit from a structured evaluation framework that assesses potential organizational use cases for AI across multiple dimensions simultaneously. The most *effective* frameworks consider business impact, technical feasibility, data readiness, organizational capability, and strategic alignment as distinct but interrelated criteria. Business impact assessment should quantify the financial opportunity through measures like revenue enhancement, cost reduction, or risk avoidance, and estimate the probability of realizing that value given implementation realities. Technical feasibility assessment should evaluate whether sufficient modeling approaches exist and whether the problem structure is amenable to AI-based solutions. Data readiness evaluation should assess whether the necessary training data exists, is accessible, and meets quality thresholds required for reliable model development.

A retail enterprise, for example, may find that predictive inventory management and dynamic pricing together account for the majority portion of potential revenue uplift available from AI, while more speculative applications (e.g., automated in-store visual merchandising) offer limited near-term value. Healthcare systems often prioritize AI-assisted diagnostic imaging and predictive readmission modeling due to their high clinical impact and ease of implementation. Conversely, while ambient clinical documentation is promising, its adoption is currently hindered by a lack of seamless integration with existing electronic health record systems. By applying a disciplined evaluation framework, organizations can concentrate resources on initiatives with strong economic justification and avoid the dilution of effort that characterizes unfocused AI programs.

Several common traps undermine business use case prioritization. The *novelty bias* leads organizations to favor technically sophisticated or conceptually exciting applications, such as generative AI content creation or autonomous decision-making, over simpler but higher-value opportunities such as anomaly detection or demand forecasting. The *feasibility trap* leads teams to favor projects that are technically straightforward but strategically marginal. The *consensus trap* leads cross-functional committees to select use cases that satisfy multiple constituencies but lack the focused impact needed to drive material outcomes. Effective prioritization requires explicit mechanisms to counteract these biases, including rigorous financial modeling, external benchmarking, and executive accountability for prioritization decisions.

Business organizations should also distinguish between use cases that serve as capability-building exercises and those designed for near-term value delivery. Early-stage organizations may need to invest in foundational use cases, such as data quality improvement or model monitoring infrastructure, before high-value applications become technically viable. A strategy that sequences these investments deliberately, treating infrastructure development as a prerequisite for value creation rather than an alternative to it, is more likely to produce sustained impact than one that attempts to shortcut foundational work in pursuit of faster returns.

The Build-Buy-Partner Decision Framework

Once promising business use cases have been identified and prioritized, organizations must evaluate how best to develop the AI capabilities required to address them. The classical build-versus-buy-versus-partner question is central to this evaluation, and the answer has significant implications for competitive positioning, capital allocation, talent requirements, and time-to-value.

Building AI capabilities internally offers maximum control, differentiation, and intellectual property ownership. Organizations that *build* develop proprietary data assets, domain-specific models, and institutional expertise that competitors cannot easily replicate. This pathway is particularly appropriate when AI is central to competitive differentiation (e.g., when the organization's AI capability is itself a source of strategic advantage) and when the problem domain is sufficiently specialized that commercial solutions cannot adequately address it. The trade-offs are substantial: internal development requires significant upfront investment in talent, infrastructure, and lifecycle management, and the time required to develop, validate, and deploy high-quality models can be considerable.

Buying commercial AI products or services provides speed and operational simplicity. The market for enterprise AI solutions has matured substantially, with credible commercial offerings available across domains including natural language processing, computer vision, predictive analytics, and recommendation systems. For business use cases where speed-to-deployment is paramount and where competitive differentiation does not depend on AI-specific capability, commercial solutions can deliver meaningful value without the complexity of internal development. The principal *risk* is vendor dependency: organizations that rely heavily on commercial AI solutions may find their strategic flexibility constrained as pricing models evolve, as vendors prioritize competing customers, or as solution capabilities fall short of emerging requirements.

Partnerships represent a pragmatic middle path, blending internal knowledge with external expertise. Common partnership models include collaborations with cloud "hyperscalers" such as AWS, Microsoft Azure, and Google Cloud that provide foundational AI infrastructure and pre-trained models; collaborations with specialized AI firms that offer domain-specific capabilities; and research partnerships with academic institutions that provide access to frontier modeling techniques. Partnerships can accelerate capability development while preserving some degree of internal control, but they require careful governance to manage intellectual property rights, data access obligations, and strategic alignment over time.

The most appropriate choice across these pathways is *context-dependent*, and most organizations will employ a portfolio of approaches rather than a single strategy. A useful decision architecture evaluates each business use case across four dimensions: *competitive centrality* (how central is this AI capability to the organization's competitive differentiation?), *organizational readiness* (does the organization have or can it develop the talent and infrastructure required for internal development?), *time sensitivity* (how urgent is deployment, and what is the cost of delayed implementation?), and *data proprietary advantage* (does the organization possess unique data assets that would give internally developed models a material advantage over commercially available alternatives?).

Business use cases that score high on competitive centrality and proprietary data advantage are strong candidates for internal development, even if the investment required is substantial. Use cases that are time-sensitive and where competitive differentiation is modest are better suited to commercial solutions. Use cases that require specialized technical expertise unavailable internally, but where some degree of customization is needed, may benefit from structured partnerships. Organizations that apply this framework systematically will develop a cleaner logic for capital allocation and make more consistent build-buy-partner decisions across their portfolio.

As the AI vendor landscape continues to grow in complexity, encompassing foundation model providers, application layer vendors, infrastructure specialists, and professional services firms, business organizations face the additional challenge of managing an increasingly complex ecosystem of external partners. Effective ecosystem management requires establishing clear criteria for vendor selection and evaluation, maintaining architectural flexibility to avoid excessive dependency on any single provider, and building internal capabilities sufficient to evaluate, integrate, and govern externally sourced AI components. Organizations that develop this ecosystem management capability will be better positioned to take advantage of rapidly evolving AI capabilities while managing the risks associated with vendor concentration.

As business organizations progress in their strategy development, many discover that AI enables entirely new business models and revenue architectures that extend well beyond the automation of existing business processes. Unlike traditional digital technologies, AI does not merely accelerate or streamline what organizations already do; it can fundamentally reshape how value is created, delivered, and captured. Recognizing and exploiting these business model opportunities is one of the highest-value applications of strategic AI thinking.

The conceptual shift is significant! Process automation asks: how can we do what we currently do more efficiently? Business model innovation asks: what could we do that was previously economically infeasible, technically impossible, or organizationally impractical and what would be the market value it provides? AI enables business model innovation by dramatically reducing the cost of prediction, by enabling personalization at scale, by making real-time optimization tractable across complex systems, and by creating the foundation for outcome-based rather than activity-based commercial relationships.

Several business model archetypes have emerged as particularly powerful applications of AI capability. *One major archetype* is outcome-based pricing models, enabled by predictive analytics, allow organizations to charge for guaranteed results rather than products or services. Industrial manufacturers, as an example, are increasingly offering guaranteed uptime contracts for complex equipment (charging for availability rather than components) made economically viable by AI-powered predictive maintenance that dramatically reduces unexpected failure rates. Yet another example involves insurance providers that are developing usage-based models in which premiums are dynamically adjusted based on real-time behavioral data, creating more accurate risk pricing and stronger customer alignment.

Personalization-at-scale represents a *second major archetype*. Consumer-facing organizations can leverage AI-driven recommendation systems, dynamic pricing, and adaptive content delivery to increase customer lifetime value through subscription enhancements, personalized product recommendations, and tailored service configurations. The economic logic is compelling: personalization reduces customer acquisition costs, increases retention rates, and captures a larger share of willingness to pay across diverse customer segments. Streaming platforms, e-commerce operators, and financial service providers have demonstrated the commercial power of this model at scale.

In financial services, AI-driven risk scoring and fraud detection algorithms are transforming underwriting economics and operational efficiency simultaneously. Credit models that incorporate alternative data sources, such as payment history, transaction patterns, and digital footprint data, can assess creditworthiness with greater accuracy than traditional scoring models, expanding addressable markets while managing risk more effectively. In healthcare, AI-enabled diagnostic tools and clinical decision support systems are creating the foundation for value-based care models in which providers are compensated for patient outcomes rather than services delivered.

A distinct and increasingly important category of AI-enabled business model innovation involves the monetization of data and AI-generated assets. Business organizations that possess rich, proprietary datasets may find that these assets have commercial value beyond their internal applications, either as datasets for model training, as the basis for AI-powered analytics services, or as inputs to industry-wide benchmarking platforms. Financial data providers, healthcare networks, and logistics operators are among the categories of organizations exploring these opportunities. The strategic and ethical dimensions of data monetization require careful consideration, particularly regarding customer consent, data sovereignty, and competitive implications.

Constructing and Managing an AI Initiative Portfolio

Strategic planning for AI must consider the composition of the organization's initiative portfolio holistically. Just as financial portfolios balance short-term liquidity against long-term growth, and just as innovation portfolios balance core business improvements against exploratory ventures, AI portfolios must combine immediate tactical improvements with more ambitious transformative initiatives. Managing the portfolio as a coherent whole, rather than as a collection of independently evaluated projects, allows organizations to optimize risk-adjusted returns, build capability progressively, and maintain organizational momentum through early wins while pursuing longer-horizon opportunities.

A useful organizing framework for AI portfolio construction draws on the classic *three-horizon model of innovation management*, adapted to the specific characteristics of AI deployment. Horizon *one* initiatives target near-term value delivery, typically within six to eighteen months, through the application of AI to well-understood processes with readily available data. Examples include automating labor-intensive document processing, enhancing customer service through AI-powered conversational assistants, or improving demand forecasting accuracy through machine learning models that replace statistical baselines. These projects deliver rapid financial returns, build organizational confidence in AI, and develop the foundational capabilities (e.g., data pipelines, model monitoring, change management) that support more ambitious initiatives.

Horizon *two* initiatives target medium-term value creation, typically over an eighteen-month to three-year horizon, through more substantial application of AI to core business processes. Examples include AI-augmented clinical decision support, intelligent supply chain optimization that integrates multiple data streams and optimization objectives, or personalization engines that reshape customer engagement models. These initiatives require more significant investment in data infrastructure, model development, and organizational change management, but they also offer commensurately larger financial impact and strategic differentiation.

Horizon *three* initiatives target longer-term, transformative outcomes that may require three to five years or more to achieve, and that fundamentally redesign business models or competitive positioning. Examples include fully autonomous supply chain orchestration, AI-native product lines that continuously adapt to individual user preferences, or clinical AI systems capable of generating diagnostic insights previously available only to specialist physicians. These initiatives carry higher technical and organizational risk, but they represent the frontier of competitive advantage; organizations that develop strategic clarity about their horizon three ambitions are better positioned to make the intermediate investments required to reach them.

Effective portfolio management requires governance mechanisms that enable dynamic resource reallocation as initiatives develop and as the external environment evolves. Static annual portfolio reviews are *insufficient* given the pace of AI capability development and the iterative nature of model building. Leading organizations establish portfolio review cadences that occur at least quarterly, with defined criteria for escalating, descaling, accelerating, or terminating initiatives based on observed performance against technical milestones and business KPIs. These governance mechanisms should be embedded in the AI strategy itself, not treated as an afterthought.

Portfolio balance also has cultural implications. Organizations that invest exclusively in horizon one initiatives tend to develop a capability profile oriented toward operational improvement rather than strategic transformation or a posture that may be competitively adequate in the near term but leaves the organization exposed as AI-native competitors move upstream. Conversely, organizations that invest disproportionately in horizon three ambitions without sufficient near-term wins risk losing organizational commitment to AI as a strategic priority. Portfolio balance is therefore not simply a financial optimization problem but a cultural and organizational one.

AI Maturity Assessment: Understanding Organizational Readiness

Understanding an organization's current level of AI readiness is a prerequisite for effective portfolio planning and strategy development. Business organizations that attempt to pursue AI initiatives without an honest assessment of their current capabilities frequently encounter execution failures that are not attributable to flawed strategy but to inadequate foundational readiness. AI *maturity models* provide a structured means of evaluating capabilities across the dimensions that determine an organization's ability to develop, deploy, and sustain AI at scale.

Mature frameworks for AI readiness assessment typically evaluate organizations across five to seven core dimensions: data quality and accessibility; technical infrastructure and platform capability; model development practices and MLOps maturity; organizational governance and risk management structures; talent depth and capability breadth; cultural readiness and change management capacity; and executive leadership and strategic alignment. Each dimension can be assessed on a progression from foundational capability (ad hoc, undocumented, reactive) through developing capability (structured but inconsistent), to mature capability (standardized, measurable, proactive), to leading capability (optimized, adaptive, and a source of competitive advantage).

Maturity assessments serve multiple strategic functions. They enable organizations to benchmark themselves against industry peers and identify relative capability gaps. They reveal bottlenecks that may impede progress. For example, an organization may discover that although it has strong data science talent, its data infrastructure lacks the robustness necessary for reliable production deployment, or that its governance structures are insufficient for managing regulatory obligations in sensitive application domains. They provide the empirical basis for sequencing investments more effectively, addressing foundational gaps before pursuing advanced applications.

A common finding in maturity assessments is that data readiness lags technical capability. Many organizations have invested in data science talent and analytical tools but have not made the complementary investments in data engineering, data governance, and data platform architecture that reliable AI deployment requires. Models trained on poor-quality, inconsistently labeled, or inadequately governed data will produce unreliable outputs regardless of the sophistication of the modeling approaches employed. Maturity assessment surfaces these gaps explicitly, allowing organizations to sequence investments in data infrastructure as a strategic priority rather than treating it as a technical detail.

A particularly important conceptual distinction that maturity assessment often surfaces is the difference between digital transformation and AI-native transformation. *Digital transformation* tends to focus on automating or streamlining existing processes through the introduction of new software tools, cloud migration, or workflow digitization. *AI-native transformation* begins with a more ambitious assumption: that processes, products, and business models can be fundamentally redesigned using predictive, generative, or autonomous capabilities, rather than merely accelerated or digitized.

The difference is not merely technical but conceptual and organizational. A digitally transformed bank may deploy robotic process automation to accelerate loan processing workflows, thereby reducing processing time and operational costs. An AI-native bank uses predictive analytics to personalize credit decisions at the individual level, redesign risk models that incorporate non-traditional data sources, and create new forms of customer engagement that anticipate financial needs before customers articulate them. A digitally transformed retailer may deploy e-commerce infrastructure and inventory management software. An AI-native retailer builds a continuously learning demand intelligence system that dynamically adjusts pricing, inventory positioning, and promotional strategies in response to real-time signals.

Understanding this distinction allows leaders to set appropriate expectations regarding investment requirements, implementation timelines, organizational change complexity, and competitive impact. Business organizations that mistake digital transformation for AI-native transformation tend to underestimate the investment required, overestimate the speed of impact, and fail to develop the foundational data and modeling capabilities that AI-native transformation demands.

Effective AI strategy development is not a task that can be delegated to any single function. It requires the active participation of executives, data scientists, engineers, operations teams, compliance officers, and domain experts who collectively possess the knowledge needed to identify where AI can add value, assess the feasibility of proposed solutions, and ensure that implementation plans reflect both strategic ambition and operational reality. Business organizations that develop AI strategies within functional silos (with the technology function leading independently of business units, or with business units developing use cases without adequate technical input) consistently underperform those that invest in structured cross-functional discovery processes.

The rationale for cross-functional collaboration is both epistemic and political. Epistemically, the identification of high-value AI opportunities requires knowledge that is distributed across the organization: domain experts understand which decisions are most consequential and most difficult to make well; data engineers understand what data is available and what its quality characteristics are; compliance officers understand which regulatory constraints apply; operations teams understand which processes are actually executed versus theoretically designed. No single function possesses "all" of this knowledge. Politically, initiatives that are developed with broad cross-functional participation tend to secure stronger organizational commitment, encounter fewer implementation obstacles, and achieve faster adoption than those developed within functional silos.

Structured Discovery Methodologies

Leading organizations employ structured discovery methodologies that systematically map organizational pain points and opportunities to potential AI solutions. Common approaches include AI opportunity workshops, in which cross-functional teams collaboratively identify and evaluate potential business use cases against a standardized set of criteria; process mapping exercises, in which high-volume, decision-intensive processes are analyzed in detail to identify where AI-driven decision support or automation would be most valuable; and outside-in reviews, in which industry benchmarking and competitive analysis are used to identify applications that peer organizations have deployed successfully and that may represent strategic gaps.

In practice, business organizations often concentrate initial discovery efforts on complex, high-volume processes such as fraud detection and financial crime management, predictive equipment maintenance and asset reliability, personalized customer engagement and next-best-action recommendation, supply chain planning and demand sensing, and clinical decision support and care management. The viability of each opportunity depends on the availability of relevant training data, the stability and observability of the underlying process, the organization's capacity to implement and maintain AI-driven workflows, and the existence of clear feedback loops that enable model performance monitoring and improvement.

A critical discipline in cross-functional discovery is the translation of opportunity identification into rigorous business cases that support investment decisions. Business use cases for AI initiatives should quantify the expected financial impact under realistic assumptions, document the data and infrastructure requirements, estimate implementation costs and timelines, assess technical and organizational risks, and define the key performance indicators against which initiative success will be measured. Business use cases that rely on optimistic assumptions about data quality, implementation complexity, or organizational adoption tend to produce disappointing outcomes. Organizations should build systematic challenge processes into their business use case development including independent review by parties without a stake in the outcome to ensure that investment decisions are made on a realistic basis.

Metrics and performance measurement provide an essential dimension of a credible AI strategy. Without rigorous measurement frameworks, organizations cannot determine whether AI initiatives are delivering the value anticipated, cannot make informed decisions about resource reallocation, and cannot build the organizational confidence in AI that sustained investment requires. Yet measurement in AI is genuinely complex: the metrics that are easiest to produce, such as model accuracy, precision, recall, latency, and other technical performance indicators, rarely capture the business impact that ultimately justifies AI investment.

A successful AI strategy reframes evaluation criteria in terms of financial outcomes, customer value, risk mitigation, and operational efficiency. The strategy articulates clear key performance indicators (KPIs) that correspond to the organization's strategic objectives, ensuring that AI initiatives are accountable to economic and operational results rather than abstract technical benchmarks. This reframing requires deliberate organizational design: who is responsible for measuring business impact? How are model performance metrics connected to operational outcomes? What governance mechanisms ensure that measurement results influence resource allocation decisions?

Effective measurement architectures for AI initiatives typically operate at multiple levels simultaneously. At the model level, technical performance metrics such as accuracy, calibration, fairness, robustness, and drift all provide insight into whether the underlying AI system is behaving as designed. At the process level, operational metrics such as decision quality, cycle time reduction, error rates, and cost per transaction serve to capture whether the AI system is improving the specific process it was designed to address. At the business level, financial and strategic metrics such as revenue impact, cost reduction, customer retention, risk-adjusted return will assess whether process improvements translate into organizational value.

A common failure event is created by the disconnection between model-level and business-level metrics. An AI system may achieve high model accuracy while failing to produce business impact. For example, because the process it supports was not a significant driver of business outcomes, because adoption was insufficient to realize the expected scale benefits, or because the model's recommendations were systematically overridden by human decision-makers, business impact may be negated. Measurement architectures that span all three levels provide earlier warning of these disconnections and enable more effective corrective action.

Beyond individual initiative measurement, organizations need mechanisms for assessing the aggregate return of their AI portfolio. Portfolio-level measurement should track the total value delivered by AI initiatives across the organization, expressed in financial terms that are comparable to other capital investment categories. This result requires establishing consistent methodologies for attributing financial impact to AI interventions such as a methodological challenge, since AI systems typically operate as one factor among many in complex business processes. Business organizations that invest in developing rigorous impact attribution methodologies will be better positioned to make the case for continued AI investment and to allocate resources effectively across competing opportunities.

Talent Strategy, Capability Development, and Organizational Culture

Talent and capability planning are among the most consequential (and most frequently underestimated) dimensions of AI strategy. The effectiveness of any AI initiative ultimately depends on the people who design, build, deploy, govern, and use AI systems. Organizations must evaluate the depth and breadth of expertise required across the full range of roles that AI deployment demands, including data engineering, statistical modeling and machine learning, MLOps and LLMOps engineering, responsible AI oversight, and translational roles that bridge technical capabilities with business objectives.

The AI talent market is intensely competitive. The supply of experienced data scientists, ML engineers, and AI architects has grown significantly in recent years; but demand has grown faster, particularly for talent with experience in production deployment, large language model applications, and responsible AI implementation. Organizations that rely exclusively on external hiring to meet AI talent needs will face sustained pressure on compensation, availability, and retention. A balanced approach such as combining targeted external hiring for specialist roles with systematic internal capability development is more sustainable and often more effective.

Internal capability development for AI should be structured across multiple levels of the organization. For technical practitioners, this action includes formal training in modern machine learning methods, MLOps practices, and AI governance frameworks. For business leaders and domain experts, AI literacy programs that develop sufficient technical understanding to participate meaningfully in business use case evaluation, requirements definition, and performance assessment are essential. For executives, strategic AI education that connects technical capabilities to business model implications and competitive dynamics enables more informed strategic decision-making.

Several organizational archetypes have now emerged for structuring AI capability development. The centralized AI *Center of Excellence* model concentrates AI talent in a shared services function that works with business units on a project basis, an approach that facilitates knowledge sharing and maintains consistent practices but can create bottlenecks and reduce business unit ownership of AI outcomes. The *federated* model embeds AI capability within individual business units, promoting domain expertise and business alignment but potentially creating inconsistent practices and duplication of foundational capability. A *hybrid* model, one that combines a central platform and governance function with embedded business unit capability, is increasingly common among organizations that have reached meaningful AI maturity.

Organizational culture plays a decisive role in AI strategy outcomes and represents a role that is frequently acknowledged but insufficiently addressed in strategy development processes. Successful AI strategies depend on cultivating organizational cultures characterized by data-driven decision-making, intellectual curiosity, tolerance for experimentation, and cross-functional collaboration. Resistance to change, siloed organizational structures, and risk-averse cultures can impede AI progress even when technical capabilities appear strong. Executive business and technology leaders who articulate clear cultural expectations, model data-informed decision-making in their own practice, and create visible consequences for reverting to intuition-based decision-making in domains where AI provides superior guidance, tend to accelerate cultural transformation.

The relationship between AI adoption and workforce dynamics also requires explicit strategic attention. AI systems that automate or augment human tasks create legitimate concerns about job security, skill obsolescence, and organizational equity that, if unaddressed, generate resistance that undermines adoption. Business organizations that communicate transparently about how AI will change roles, invest in reskilling and upskilling programs, and demonstrate genuine commitment to managing workforce transitions responsibly are better positioned to sustain organizational commitment to AI transformation over the multi-year timescales that meaningful impact requires.

Data Architecture, Infrastructure, and the AI Technology Stack

No AI strategy can succeed without a rigorous plan for data acquisition, management, and governance. Data is the *raw material* from which AI systems learn; the quality, quantity, relevance, and governance of available data largely determine the ceiling on AI system performance. Business organizations that invest in AI modeling capability without making commensurate investments in data infrastructure consistently encounter deployment failures, model reliability problems, and regulatory compliance challenges that limit the value their AI initiatives can produce.

A data strategy in support of AI pursuits should address several core capabilities: data acquisition and sourcing (what data does the organization need, from which sources, and under what terms?); data engineering and integration (how is raw data transformed into high-quality, consistently structured datasets suitable for model training?); data quality management (what processes ensure that data meets accuracy, completeness, consistency, and timeliness requirements?); data governance (who owns data, who can access it, under what conditions, and how is compliance with privacy regulations maintained?); and data platform architecture (what storage, processing, and serving infrastructure enables AI systems to access the data they need, at the scale and latency required?).

The effectiveness of AI initiatives also depends on whether the organization has the technical infrastructure necessary to support the full lifecycle of AI system development and operation. The AI technology stack encompasses data ingestion and processing pipelines; feature stores that maintain curated, reusable feature sets for model training; model training and experimentation platforms; model registries that track model versions, metadata, and lineage; model serving infrastructure that delivers inference at the required scale and latency; monitoring systems that detect model performance degradation, data drift, and fairness violations; and governance frameworks that document model provenance, track decision outcomes, and support regulatory compliance.

The emergence of *MLOps* and *LLMOps* as disciplined engineering practices reflects the recognition that deploying AI at enterprise scale requires the same engineering rigor applied to other software systems. Models that perform well in development frequently degrade in production as input data distributions shift, as business processes evolve, or as the phenomena being modeled change. Operationally mature AI organizations treat model monitoring and retraining as continuous engineering disciplines, maintaining the infrastructure and processes needed to detect and respond to model performance issues before they produce significant business impact.

The architectural choices between cloud, on-premises, and hybrid AI infrastructure have significant implications for cost, flexibility, security, and regulatory compliance. Cloud platforms offered by AWS, Microsoft Azure, and Google Cloud provide access to powerful AI infrastructure (including GPU computing, managed ML platforms, and foundation model APIs) without the capital investment required for on-premises deployment. For most organizations, cloud infrastructure represents the most practical starting point for AI experimentation and development. However, data sovereignty requirements, latency constraints, regulatory obligations, and cost considerations may justify on-premises or hybrid architectures for specific workloads. AI strategy should include explicit infrastructure architecture principles that govern these decisions consistently across the organization, rather than allowing ad hoc choices to accumulate into an unmanageable architectural complexity.

From Strategy to Implementation: Building an AI Roadmap

A well-constructed AI strategy must be translated into an *operational roadmap* that defines the sequence of investments, capability developments, and organizational changes required to achieve strategic objectives. The roadmap serves as the primary vehicle for strategy execution, converting the aspirational language of strategic planning into concrete initiatives, resource allocations, timelines, and accountability assignments that guide organizational action.

An effective AI roadmap is organized around multiple time horizons and capability dimensions simultaneously. It sequences capability-building investments such as data infrastructure, platform development, talent acquisition, and governance framework establishment all as prerequisites for value-creation initiatives. A roadmap prioritizes initiatives based on their expected business impact, technical feasibility, and organizational readiness.

Also, a roadmap defines explicit milestones and decision points at which progress will be assessed and portfolio allocations will be reviewed. And it identifies the organizational changes (e.g., new roles, revised processes, updated governance structures, cultural interventions) that are required to support successful implementation.

The implementation of an AI strategy is fundamentally an organizational change management challenge, not merely a technical one. AI systems that are technically sound but organizationally unacceptable (because affected employees do not understand them, do not trust them, or do not feel that their concerns have been heard) fail to deliver their intended value. Change management planning should begin at the strategy development stage, not at the implementation stage, with explicit attention to stakeholder identification and engagement, communication planning, training and capability development, and feedback mechanisms that enable the strategy to be refined based on implementation experience.

Executive sponsorship is one of the most significant determinants of AI implementation success. AI initiatives that have visible, sustained, and credible executive champions (and who communicate the strategic rationale for AI investment, model data-driven decision-making in their own practice, and hold the organization accountable for AI-related commitments) achieve higher adoption rates and greater business impact than those that lack this level of leadership engagement. Building executive sponsorship into the roadmap as an explicit deliverable, with defined roles and responsibilities for specific executives, is a practical mechanism for ensuring that leadership commitment is maintained through the implementation period.

Finally, AI strategy should be understood as a living discipline rather than a static plan. The rapid pace of AI capability development, the evolution of the competitive landscape, and the organizational learning that accumulates through implementation experience all create continuous pressure to reassess and refine strategic priorities. Organizations that establish regular strategy review processes, with defined triggers for out-of-cycle reassessment, such as significant model capability announcements, competitive moves, or regulatory developments, will be better positioned to maintain strategic relevance as the AI landscape evolves.

The most sophisticated AI strategies incorporate feedback loops that enable learning from implementation experience to inform strategic refinement. What business use cases proved more or less tractable than anticipated? What organizational capabilities proved to be binding constraints? What business model innovations generated unexpected customer response? What governance challenges emerged in production deployment that were not anticipated in the planning process? Institutionalizing these learning processes through structured retrospectives, performance reviews, and strategic planning cycles transforms AI strategy from a one-time exercise into an organizational capability that compounds in value over time.

Developing an enterprise AI strategy requires a holistic understanding of organizational priorities, technological capabilities, market dynamics, and ethical responsibilities. It encompasses the identification and prioritization of high-value opportunities; the evaluation of build-buy-partner pathways for capability development; the design of AI-enabled business models and revenue architectures; the construction of a balanced initiative portfolio across multiple time horizons; the rigorous assessment of organizational maturity and readiness; the cultivation of cross-functional talent, culture, and discovery processes; the integration of measurement frameworks that connect technical performance to business outcomes; and the embedding of governance, ethics, and risk management principles throughout the strategy lifecycle.

The distinction between organizations that develop effective AI strategies (and those that do not) is not primarily a function of technical capability or financial resources. It is a function of strategic discipline or the willingness to make explicit choices about where to concentrate investment, the organizational courage to deprioritize attractive-seeming opportunities that lack strong strategic justification, and the leadership commitment to maintain organizational focus through the multi-year implementation timelines that meaningful AI transformation demands.

When executed thoughtfully, an *AI strategy* enables organizations to transcend isolated experiments and achieve meaningful, measurable transformation. Moreover, it provides the coordinated structure needed to turn technological promise into economic performance thus ensuring that AI becomes a durable source of competitive advantage rather than an aspirational concept. For business and technology leaders and strategists who develop this capability, AI strategy is not simply a planning exercise. AI strategy planning is the discipline through which organizations define and secure their competitive position in an economy increasingly shaped by artificial intelligence.

CHAPTER 5

The Economics and Valuation of AI

THE RAPID ADOPTION OF AI across industries has introduced a new economic logic into modern enterprise strategy. Unlike traditional information technology investments, those historically characterized by heavy upfront capital expenditure and multi-year depreciation cycles, AI initiatives often rely on flexible, consumption-based models that convert what was once capital-intensive infrastructure into variable operational expenditure. This structural shift has altered how organizations plan, evaluate, and scale technology-driven capabilities. As the economic mechanisms underpinning AI differ fundamentally from those of legacy systems, executives must understand these dynamics to ensure that AI deployment aligns with broader financial and strategic objectives.

CAPEX to OPEX

Central to this transformation is the emergence of cloud-based AI platforms that enable organizations to access computing power, storage, and advanced machine learning capabilities without purchasing or maintaining dedicated physical infrastructure. Cloud providers have introduced scalable architectures in which resources can be provisioned on demand, allowing firms to pursue multiple AI experiments with minimal upfront commitment. The conversion from capital expenditure (capex) to operating expenditure (opex) lowers barriers to entry, reduces financial risk associated with technology obsolescence, and encourages experimentation. For large enterprises, this model supports rapid scaling of successful initiatives, while for startups it provides access to high-performance computing previously reserved for well-capitalized incumbents. The result is a more fluid innovation environment in which capital allocation becomes tied directly to demonstrated value.

Within this broader opex framework, token- and inference-based pricing models further refine the economics of AI usage. Providers of LLMs and other advanced AI services increasingly charge based on consumption (such as the number of tokens processed, inference calls made, or model outputs generated) rather than fixed licensing fees. This granular pricing structure aligns cost with activity, introduces a high degree of transparency, and allows organizations to associate AI expenses directly with specific business processes or customer interactions. A financial services company deploying an LLM for real-time client communications, for example, pays only for the volume of queries processed, enabling precise cost attribution and minimizing idle capacity. Such models reward efficient application design and encourage firms to optimize prompts, workflows, and model configurations to contain costs while preserving performance.

Evaluating the economic value of AI projects requires analytical tools capable of capturing uncertainty and strategic flexibility. Traditional discounted cash flow (DCF) techniques provide a baseline for assessing expected financial returns, but they often prove insufficient in contexts where outcomes are highly uncertain, volatile, or path-dependent. AI initiatives frequently possess significant optionality: early-stage pilots may serve as gateways to enterprise-wide adoption, multiproduct expansion, or new revenue models if successful. Real options analysis provides a means of quantifying these latent opportunities by valuing the managerial flexibility to scale, pivot, or abandon projects as new information emerges. A predictive maintenance system piloted in one facility, for example, may ultimately be deployed across dozens of plants worldwide, generating returns far beyond the initial scope. Incorporating real options into valuation frameworks ensures that decision-makers recognize the strategic upside embedded in AI investments rather than undervaluing them by focusing exclusively on static projections.

At the same time, the inherent uncertainties associated with AI technologies necessitate adjustments to standard financial assumptions. Variability in model performance, data quality, regulatory exposure, user adoption, market acceptance, and competitive response introduces layers of risk that must be reflected in valuation models. Finance teams increasingly rely on scenario analysis, sensitivity tests, "Monte Carlo" simulations, and probabilistic modeling to quantify the potential range of outcomes. These approaches allow organizations to incorporate volatility explicitly, stress-test assumptions, and understand how external or internal shocks might alter projected returns. By embracing probabilistic rather than deterministic models, organizations enhance the robustness of capital allocation decisions.

Intangible Capital: The Role of Training and Organizational Redesign in AI ROI

When organizations attempt to measure the return on their AI investments, they almost universally begin in the wrong place. They build discounted cash flow models around projected cost reductions, construct ROI frameworks anchored to headcount displacement calculations, and present boards with revenue uplift projections derived from pilot results that rarely survive contact with enterprise-scale deployment. These approaches are not wrong in what they measure, i.e., cost reduction and revenue growth are legitimate and important dimensions of AI value, but they are profoundly incomplete in what they ignore.

The most significant and most durable sources of AI return on investment are *not* found in the operating expense line or the revenue forecast. They are found in the accumulation of intangible capital or the organizational knowledge, human capability, process intelligence, and institutional learning capacity that AI transformation builds inside the enterprise when it is executed with sufficient strategic intentionality. Intangible capital of this kind does not appear on the balance sheet under current accounting standards, cannot be easily modeled in a DCF analysis, and resists the quarterly reporting cadences that govern most executive attention. But its economic significance is not diminished by its invisibility in conventional financial frameworks. In the medium to long term, the intangible capital that an organization builds (or fails to build) during its AI transformation will be a more reliable predictor of competitive position and enterprise value than any specific model deployment or automation initiative the organization executes along the way.

The concept of intangible capital in the context of AI investment encompasses several distinct but interconnected categories of organizational asset. The *first* and most immediately recognizable is human capital or the AI literacy, technical fluency, and augmented decision-making capability that the workforce develops through sustained engagement with AI tools and through the deliberate training investments the organization makes to accelerate that development. This is not simply a matter of teaching employees to use new software. Human capital represents a fundamental upgrade in the cognitive infrastructure of the organization such as the collective capacity to frame problems in ways that AI can help solve, to critically evaluate AI outputs rather than accept them uncritically, to identify high-value AI applications within specific domain contexts, and to collaborate effectively with AI systems in ways that amplify rather than merely replicate prior human capability.

Research by economists Erik Brynjolfsson and Lorin Hitt, whose work on IT investments established the foundational framework for understanding intangible capital in technology contexts, demonstrated that organizations investing in complementary human capital alongside technology deployments consistently outperformed those treating technology as a standalone investment, and often by margins that dwarfed the direct returns from the technology itself. The same dynamic applies with even greater force to AI, where the gap between the potential of the technology and the organizational capability required to realize that potential is wider than it has been *for any prior* enterprise technology category.

The *second* major category of intangible capital created through AI transformation is organizational process capital meaning the redesigned workflows, decision frameworks, operating procedures, and institutional routines that embed AI capability into the way the organization functions rather than treating it as an add-on layer sitting alongside prior practice. This distinction matters enormously for ROI calculation purposes, and it is one that most AI investment frameworks systematically undervalue. An AI tool deployed on top of an un-redesigned workflow captures only a fraction of the value that the same tool deployed within a workflow specifically architected around its capabilities can generate.

A language model used to assist a financial analyst who still follows a report production process designed for manual research and synthesis will deliver marginal efficiency gains. The same language model deployed within a process redesigned to leverage its synthesis speed at the front end, its pattern recognition across large document sets in the middle, and the analyst's irreplaceable contextual judgment and client relationship intelligence at the output stage will deliver transformative productivity and quality improvements that compound over time as the process is continuously refined. The investment required to achieve the second outcome (e.g., process mapping, workflow redesign, change management, or iterative optimization) is real, substantial, and almost never fully accounted for in AI project budgets. But its omission from the ROI calculation does not make it less real, its omission simply makes the ROI calculation wrong, and organizations that discover this discrepancy after deployment (when actual returns fall short of projected returns despite technically successful model performance) are frequently unable to diagnose the source of the gap because they never measured the process capital dimension of the investment from the beginning.

The *third* category is perhaps the most strategically significant and the least discussed in conventional AI economics literature: organizational learning capital or the institutional capacity to learn from AI deployments, to systematically capture and distribute the insights that AI-human collaboration generates, and to continuously improve both the AI systems themselves and the human practices surrounding them. Business organizations that build robust feedback loops between AI system performance and human domain expertise (e.g., where model errors surface actionable learning, where user behavior generates ongoing training signal, or where the organization's accumulated experience with AI deployment informs each successive initiative) are building a compounding asset that grows in value with each iteration cycle.

This learning capital is what separates organizations that achieve sustained AI advantage from those that execute technically successful initial deployments and then plateau. It represents the organizational equivalent of the data flywheel that gives platform businesses their structural competitive advantage or a self-reinforcing cycle in which AI capability generates insight, insight improves capability, improved capability attracts more use, and greater use generates richer insight. Building this flywheel requires deliberate investment in the knowledge management infrastructure, the communities of practice, the after-action review disciplines, and the cross-functional learning mechanisms that most organizations currently do not have and do not account for in their AI investment planning. This flywheel is, in the most precise economic sense, a capital investment and one that generates returns over a multi-year horizon, and which should be evaluated on that basis rather than against the quarterly value realization expectations that govern most AI program reviews.

The Non-Linear Costs of AI: Why AI Investment Defies Conventional Technology Economics

One of the most consequential and consistently underestimated characteristics of enterprise AI investment is the non-linearity of its cost structure, essentially the phenomenon by which AI program costs do not scale proportionally with AI program scope, capability, or ambition, but instead grow in ways that are discontinuous, compounding, and frequently invisible until they have already materialized into budget overruns, organizational strain, and financial outcomes that bear little resemblance to the projections that secured the original investment approval. Every experienced technology executive has encountered the phenomenon of technology projects that cost more than budgeted; it is among the most durable and most thoroughly documented patterns in the history of enterprise technology investment.

But AI's cost non-linearity is qualitatively different from the cost overrun patterns of prior technology categories, not merely in degree but in kind: AI operates through mechanisms that conventional technology investment frameworks are structurally ill-equipped to anticipate, through cost categories that standard project accounting methodologies systematically under-capture, and through feedback dynamics that cause costs to accelerate precisely when organizations are most financially and organizationally committed to continuing. Understanding AI's non-linear cost structure (not as an abstract economic phenomenon but as a set of specific, identifiable, and to a meaningful extent manageable mechanisms) is one of the most practically important capabilities that AI investment decision-makers can develop, because the organizations that understand it before they commit to major AI investments are the ones best positioned to design investment programs that deliver the returns they promise rather than the disappointments that non-linear cost dynamics, left unmanaged, reliably produce.

Infrastructure Threshold Effect

The *first* and most immediately recognizable source of AI cost non-linearity is what economists call the *infrastructure threshold effect* or the pattern by which AI capability does not become available gradually as investment increases but instead becomes available in discontinuous steps that require crossing specific infrastructure investment thresholds before any meaningful capability is accessible. In conventional software deployment, a modest investment produces modest capability (e.g., a company that spends fifty thousand dollars on a business intelligence platform gets proportionally less capability than one that spends five hundred thousand dollars) but it gets something useful at both investment levels. AI infrastructure does not follow this pattern. The data infrastructure required to support enterprise AI (i.e., the data warehouse or lakehouse capable of storing and querying the data volumes that model training requires, the data pipeline infrastructure that moves data from operational systems to analytical environments with sufficient quality and consistency for AI workloads, the feature engineering capability that transforms raw data into the structured inputs that machine learning models can use, and the MLOps infrastructure that manages the model lifecycle from training through deployment through monitoring) represents a set of *fixed* costs that must be substantially incurred before any production AI system can be reliably deployed and maintained.

Business organizations that attempt to deploy AI without crossing these infrastructure thresholds discover, typically at the point of attempting to scale a successful pilot to production, that the infrastructure investment they deferred to keep the pilot cost-contained is now the rate-limiting constraint on everything they want to do with AI; the cost of building it at scale, under the time pressure of a board expecting the returns the pilot promised, is substantially higher than it would have been if it had been built deliberately and in advance. The infrastructure threshold effect means that AI investment programs face a paradox that conventional technology investment frameworks cannot easily accommodate: the investment required to do AI properly is largely fixed and largely upfront, while the returns are variable and largely deferred. This effect results in a cost-return timing mismatch that makes AI investment look more expensive and less productive in its early stages than it actually is on a lifecycle basis, and that creates exactly the short-term financial pressure that causes organizations to cut the infrastructure investment that would have made the long-term returns possible.

Data Cost Curve

The *second* source of non-linearity is the data cost curve or the pattern by which the cost of acquiring, cleaning, labeling, and maintaining the data that AI models require grows significantly faster than the quantity of data being managed, because the marginal cost of each additional unit of high-quality, well-labeled, AI-ready data is higher than the unit before it. The intuition behind this pattern is straightforward: the easiest data to collect, clean, and label is collected, cleaned, and labeled first, leaving progressively more difficult, more expensive, and more organizationally complex data problems to be solved as the program advances. A computer vision model trained to identify defects in manufactured products might be trained initially on a dataset of ten thousand well-lit, consistently framed images of clearly defective and clearly non-defective component images that are relatively easy to collect from existing quality inspection cameras and relatively easy to label because the defects are visually obvious.

Improving that model's performance in the edge cases that matter most, the marginal defects that are expensive to miss and expensive to falsely flag, the components photographed under variable lighting conditions or from non-standard angles, or the defect categories that are rare in the historical data and therefore underrepresented in the training set requires data that is dramatically more expensive per unit to acquire and label than the initial training dataset. The defects are more rare, making collection more time-consuming. The labeling requires domain expertise that is more expensive than the general labeling the initial dataset required. Also, the organizational processes required to collect, manage, and maintain the data quality of a larger and more heterogeneous dataset are more complex and more expensive to operate than those that served the initial program. This data cost curve is not a temporary feature of early-stage AI programs that resolves as the data infrastructure matures. Rather, it is a structural characteristic of AI development that reflects the fundamental economics of information quality or the principle that the last unit of data quality is always more expensive than the first, and that the AI use cases that generate the most business value are almost always the ones that require the highest data quality, the deepest domain expertise in labeling, and the most sophisticated data governance to maintain.

The Capability Expansion Cost Spiral

The *third* source of non-linearity is perhaps the most counterintuitive and the most organizationally disruptive: the *capability expansion cost spiral*, in which the deployment of a successful AI capability generates organizational demand for expanded capability that costs disproportionately more to deliver than the initial capability did. This dynamic is rooted in the economics of organizational expectation or the well-documented phenomenon by which successful technology deployments raise the bar for what the technology is expected to deliver, creating a ratchet effect in which each successful AI application generates requests for extensions, enhancements, and adjacent applications that collectively cost more than the original investment and that arrive before the original investment has generated the returns that would fund them. A natural language processing system deployed to automate the first-line response to customer service inquiries might handle seventy percent of inquiry volume effectively at an investment level that the business case supported.

The remaining thirty percent (the complex, ambiguous, emotionally charged inquiries that require genuine contextual understanding and judgment) are now more visible to the organization than they were before the AI system was deployed, because the contrast between the inquiries the system handles well and the inquiries it handles poorly creates organizational pressure to close the gap. Closing that gap and improving the system's performance on the hardest thirty percent of cases is dramatically more expensive per unit of improvement than achieving the initial seventy percent coverage was, because the hard cases are hard precisely because they involve the linguistic complexity, contextual nuance, and domain-specific knowledge that is most difficult and most expensive for AI systems to acquire. Meanwhile, the business units that observed the customer service deployment are requesting similar AI capabilities in their own domains (i.e., sales, marketing, compliance, HR) each of which requires its own data infrastructure, its own model development, its own integration with existing systems, and its own change management program. The cost of serving this expanded demand scales faster than the organizational AI infrastructure can be built to support it, creating a resource constraint that either delays the new applications until frustration with the AI program's responsiveness undermines organizational commitment, or accelerates them at a pace that compromises the quality and governance rigor that sustainable AI deployment requires.

AI Systems Operational Cost

The *fourth* source of non-linearity is the ongoing operational cost of AI systems in production, a category of expenditure that is almost universally underrepresented in AI project business cases and that grows in ways that are difficult to predict at project inception but that can, over a multi-year production lifecycle, exceed the initial development investment by a significant margin. The fundamental economic characteristic that distinguishes AI operational costs from conventional software operational costs is that *AI systems require continuous investment* simply to maintain their current level of performance; this requirement also has no direct analogue in conventional software, where a deployed system performs the function it was designed to perform without requiring ongoing investment in its core logic unless a deliberate enhancement is desired. AI models, by contrast, are trained on historical data that reflects the world as it was at a specific point in time, and the world changes in ways that progressively reduce the relevance of that historical data to the present-day decisions the model is being asked to support.

This phenomenon, model drift or model degradation, is not a failure of the model or the model development process. It is rather an inherent consequence of deploying statistical models in dynamic environments, and it is as predictable and as inevitable as the depreciation of a physical asset. Managing model drift requires continuous monitoring of model performance against current ground truth data, regular retraining cycles that incorporate new data and update the model's representation of current reality, validation and testing of retrained models before they replace the production model, and the governance and approval processes that responsible model deployment requires, all of which represent ongoing operational costs that the AI system requires for as long as it remains in production. These operational costs are non-linear in the aggregate because each AI system added to the production environment adds its own monitoring, retraining, and governance overhead, and the total operational cost of an AI portfolio grows faster than the number of systems in it because the interactions between systems, the shared infrastructure dependencies, and the governance complexity of managing multiple production AI systems simultaneously all create overhead that scales super-linearly with portfolio size.

Remediation Cost Asymmetry

The *fifth* and most organizationally complex source of AI cost non-linearity is what might be called the *remediation cost asymmetry* or the principle that the cost of correcting an AI system failure is always significantly greater than the cost of preventing it would have been, and that this asymmetry grows with the scale of deployment and the consequentiality of the decisions the AI system influences. This asymmetry operates through several mechanisms simultaneously. The *first mechanism* is the systematic nature of AI failure; as we discussed in the context of human-in-the-loop dynamics, AI systems fail consistently rather than randomly, meaning that a model error affects not one decision but all decisions made under the conditions that trigger the error, potentially generating large volumes of incorrect outcomes before the failure is detected.

The cost of remediating a systematic AI failure or identifying every affected decision, reversing or compensating the incorrect outcomes, rebuilding the trust of the customers, employees, or regulators affected, and redesigning the system and governance framework to prevent recurrence clearly is a function of the number of decisions made during the detection gap, which is itself a function of the quality of the monitoring infrastructure that the organization invested in during deployment. Business organizations that invest adequately in monitoring can detect systematic failures quickly and contain the remediation cost. Those that underinvest in monitoring discover systematic failures through external complaints, regulatory inquiries, or statistical anomalies in business performance metrics, by which time the remediation cost has compounded to a level that frequently exceeds the entire original development investment.

The *second mechanism* is the regulatory remediation cost that governance failures generate (such as the legal fees, regulatory fines, remediation program costs, and reputational damage expenses that AI governance failures produce in an increasingly regulated environment) none of which appear in AI project business cases because they are treated as low-probability tail risks rather than as predictable consequences of specific governance choices that the organization is making at the moment the project is designed.

The *third mechanism* is the organizational trust cost or the erosion of employee, customer, and stakeholder confidence in AI systems that follows a visible AI failure, which increases the change management investment required for every subsequent AI deployment and extends the adoption timeline in ways that defer the returns from future AI investments across the entire portfolio.

Understanding these five sources of non-linearity: 1) the infrastructure threshold effect, 2) the data cost curve, 3) the capability expansion spiral, 4) the ongoing operational cost of production AI, and 5) the remediation cost asymmetry *does not* make AI investment less attractive. The returns from well-designed and well-governed AI programs remain among the most compelling available to enterprises making technology investments in the current environment, and the competitive cost of not investing in AI is itself non-linear in ways that make the cost of adequate investment look modest by comparison.

What understanding AI's non-linear cost structure does is change the nature and quality of AI investment decisions from decisions made on the basis of business cases that capture the linear costs and ignore the non-linear ones, to decisions made on the basis of a complete and honest financial architecture that accounts for the full cost of doing AI well, the full cost of doing it poorly, and the full cost of not doing it entirely. The business organizations that develop this understanding earliest will design AI investment programs that are adequately resourced, appropriately governed, and financially structured to deliver the returns they promise, and they will do so not because they are more optimistic about AI's potential than their peers, but because they are more honest about what realizing that potential actually costs.

The Productivity J-Curve and Intangible Capital: The Hidden Investment That Determines Whether AI Transformation Succeeds or Fails

There is a moment in almost every serious enterprise AI transformation, a moment that arrives with uncomfortable reliability somewhere between twelve and twenty-four months after the initial investment commitment, when the financial evidence appears to contradict the strategic logic that justified the program. The technology has been deployed. The models are running in production. The organizational change management program has been executed, the training has been delivered, and the adoption metrics show that employees are using the AI tools at the rates the program governance requires. Still, the productivity improvements that the business case projected, the cost reductions that the CFO built into the forward financial plan, and the competitive advantages that the CEO described to the board are either not materializing at the expected scale, not materializing on the expected timeline, or materializing in forms that the measurement framework was not designed to capture and cannot therefore report.

This is the moment (the bottom of the productivity J-curve, the valley of the transformation trough) at which the organizational and financial pressures to reduce AI investment, slow the transformation program, or declare a strategic pivot away from the current approach are at their most intense and at which the consequences of yielding to those pressures are most severe. This is also the moment at which the concept of *intangible capital* becomes most analytically important because understanding what is actually happening at the bottom of the J-curve, namely what the organization is building during the period when its financial statements suggest it is losing ground, and why that building process is the prerequisite rather than the alternative to the productivity gains the program was designed to deliver, is the difference between leadership that maintains the investment discipline required to reach the inflection point versus leadership that retreats at precisely the moment when persistence would have been most valuable.

Applying the J-curve

The J-curve framework, as developed in the economics literature on general purpose technology adoption, productivity, and most rigorously applied to enterprise contexts by Erik Brynjolfsson and his collaborators at the Massachusetts Institute of Technology (MIT), describes the characteristic return profile of transformative technology investments with a precision that makes it not merely descriptively accurate but genuinely predictive and a tool that allows organizations to anticipate the shape of their AI transformation's financial trajectory before they enter it rather than being surprised by it after the fact. The curve's defining feature is the temporal separation between the cost realization and the value realization of transformative technology investment: costs are incurred early and continuously throughout the transformation program, while the productivity gains that justify those costs are deferred until the organizational complementary investments (e.g., the process redesign, the capability development, the cultural adaptation, and the structural reorganization that transformative technology requires) have been sufficiently completed to allow the technology's potential to express itself in measurable operational performance.

This temporal separation is not a design flaw in AI transformation programs that better planning could eliminate. It is a structural characteristic of all general purpose technology adoption that reflects a fundamental economic reality: technology that is capable of transforming how an organization works cannot deliver that transformation until the organization's workflows, capabilities, and structures have been redesigned around the technology's capabilities, essentially a redesign process that takes time, consumes resources, and produces measurable disruption before it produces measurable improvement. The organizations that understand this reality before they enter the transformation are the ones that design their investment programs, governance frameworks, and stakeholder communication strategies to manage through the J-curve rather than being managed by it. Those that do not understand it discover it experientially, at the bottom of the curve, under conditions of maximum financial and organizational stress that make clear-headed strategic decision-making uniquely difficult.

Hidden Costs That Drive the Initial Productivity Dip

The downward stroke of the J before the inflection point is reached are hidden not because they are genuinely invisible to careful analysis but because they fall outside the cost categories that conventional technology investment frameworks are designed to capture and that standard project accounting methodologies are structured to track. The *first and most significant* hidden cost category is process redesign investment or the organizational work required to rebuild the workflows, decision processes, operating procedures, and institutional routines that must change for AI capability to generate genuine productivity improvement rather than adding a technology layer on top of unchanged organizational processes that neutralize its impact. Process redesign of the depth that transformative AI adoption requires is not a project management exercise. It is an organizational development discipline that requires deep understanding of how work actually flows through the organization and not simply how the process documentation says it flows, but how it actually flows, through the informal workarounds, the undocumented exceptions, and the relationship-dependent coordination mechanisms that real organizations develop over time to compensate for the inadequacies of their formally documented processes.

Mapping this actual process reality, identifying the specific points at which AI capability can be most productively integrated, redesigning the workflow architecture around those integration points, and managing the organizational change required to move from existing to redesigned processes is work that consumes significant time and organizational attention (typically six to eighteen months for major process domains) and that generates no directly measurable productivity output during the period of its execution. It appears in the financial statements as cost without corresponding benefit, contributing directly to the J-curve's downward stroke while building the process capital that the upward stroke requires.

The *second hidden cost* category is human capability development investment or the comprehensive, sustained, and organizationally embedded learning infrastructure required to build the AI fluency, the human-AI collaboration skills, and the domain-specific AI application capability that genuine productivity improvement demands. The distinction between this investment and the compliance training that most organizations budget for is not only quantitative but qualitative: compliance training produces documented completion rates; genuine capability development produces changed behavior, improved judgment, and the kind of confident, creative engagement with AI tools that allows employees to use them in ways that generate productivity improvement rather than merely satisfying the adoption metrics that program governance tracks.

Building genuine AI capability across a large workforce requires sustained investment in learning infrastructure, in other words dedicated learning time that is protected from the operational demands that would otherwise consume it, coaching and mentoring relationships that provide the personalized support that self-directed learning cannot, communities of practice that connect AI practitioners across the organization in ways that accelerate collective learning, and the iterative, project-based learning experiences that develop the applied capability that conceptual training alone cannot build. This investment is expensive in time, management attention, and direct financial cost, and its returns are deferred and indirect, appearing not as measurable productivity improvements in the period of the investment but as the human capital foundation on which the productivity improvements of the J-curve's upward stroke are built. Organizations that substitute cheaper, faster, and more measurable compliance training for genuine capability development investment are not reducing the cost of AI transformation. They are reducing the probability that the transformation will deliver the returns that justify its cost, while maintaining the appearance of responsible investment management that the compliance training metrics provide.

The *third hidden cost category* is cultural adaptation investment or the organizational development work required to shift the beliefs, behavioral norms, social dynamics, and institutional identities that determine how an organization's members collectively respond to the changes that AI transformation requires of them. Culture is the most expensive and most time-consuming dimension of organizational transformation, and it is the dimension that is most systematically excluded from AI investment frameworks because it is the dimension most resistant to the quantification that investment frameworks require. The cultural shifts that AI transformation demands are not superficial, e.g., they are not resolved by a town hall presentation from the CEO, a set of revised values statements on the corporate intranet, or even a well-designed change management communication campaign. They require the genuine renegotiation of the implicit agreements that govern organizational life: agreements about what constitutes valued expertise, about how status and influence are earned and maintained, about what risks are acceptable to take and who bears the consequences when they materialize, and about the fundamental relationship between human judgment and algorithmic recommendation in the organizational decision-making processes that determine outcomes. Renegotiating these agreements requires sustained leadership attention, genuine organizational dialogue that goes beyond communication to encompass genuine listening and genuine response, and the visible behavioral changes from senior leaders that signal the seriousness of the cultural transformation more credibly than any communication campaign. Each of these requirements consumes organizational resources such as leadership time, management attention, and employee cognitive bandwidth that are genuinely scarce and that are consumed during the transformation period at the expense of the operational focus that short-term productivity maintenance requires, contributing directly to the productivity dip that characterizes the J-curve's downward stroke.

The *fourth hidden cost category* is what organizational economists call *coordination cost inflation* or the temporary but substantial increase in the organizational effort required to coordinate work during the transition period when some processes have been redesigned around AI capability and others have not, when some employees have developed meaningful AI fluency and others are still developing it, and when the formal authority structures and informal coordination mechanisms of the organization are being restructured around new roles, new workflows, and new decision rights that have not yet achieved the "taken-for-granted" quality that makes organizational coordination efficient.

Every business organization that has navigated a major operational transformation (e.g., the introduction of enterprise resource planning systems, the shift to agile development methodologies, or the transition to remote and hybrid work) has experienced this coordination cost inflation… the period during which the old ways of working have been disrupted but the new ways have not yet become habitual, and the organizational energy that habitual coordination processes require almost no conscious attention to sustain is suddenly being consumed by the deliberate, effortful, and error-prone process of figuring out how to work in a new way. In AI transformation, this coordination cost inflation is particularly pronounced because the transformation is not confined to a single process domain or organizational function but extends, in genuinely ambitious AI programs, across multiple functions simultaneously thus creating a period of organization-wide coordination stress that the financial statements capture as reduced productivity and that leadership must interpret as the organizational metabolism of genuine transformation rather than as evidence of program failure.

Reframing Hidden Costs as Intangible Capital Investments

The concept of intangible capital provides the analytical framework that transforms these hidden costs from sources of financial anxiety into comprehensible and manageable dimensions of a coherent investment strategy. Intangible capital, as introduced earlier in this chapter, refers to the organizational assets (i.e., the process architectures, human capabilities, cultural dispositions, and institutional knowledge) that AI transformation builds during the J-curve's downward stroke and that generate the productivity returns of its upward stroke. The critical analytical insight that the intangible capital framework provides is the recognition that the hidden costs described above are not costs in the conventional sense of expenditures that consume value without creating it. They are capital investments such as expenditures that create organizational assets whose value is real, durable, and compounding, even when those assets do not appear on the balance sheet and cannot be directly measured by the financial metrics that govern most investment evaluations.

A workforce that has genuinely developed AI fluency is more productive than one that has not, in ways that compound over time as AI capability continues to advance and the fluent workforce can leverage each new capability more quickly and more creatively than one that is still developing foundational understanding. A set of business processes genuinely redesigned around AI capability is more efficient and more adaptive than one that has AI layered on top of unchanged workflows, in ways that generate compounding returns as process refinement continues and AI capability improves. An organizational culture that has genuinely adapted to human-AI collaboration is more innovative and more resilient than one still negotiating the terms of that adaptation, in ways that are visible in product development speed, customer experience quality, and talent attraction and retention ...all of which have measurable financial consequences even when they resist direct attribution to the cultural transformation that produced them.

Governing AI Transformation: Metrics, Investment Discipline, and J-Curve Time Horizons

The investment governance implications of the J-curve and intangible capital framework are specific, practical, and in some respects counterintuitive relative to the instincts that conventional financial management produces. The *first implication* is that the measurement framework governing AI transformation programs must be expanded beyond financial return metrics to include leading indicators of intangible capital accumulation or metrics that allow leadership to assess whether the organization is building the process, human, and cultural capital that the J-curve's upward stroke requires, rather than waiting for the financial returns that intangible capital generates to appear in the income statement.

Process redesign completion quality assessments, AI literacy development metrics that measure behavioral change rather than training completion, cultural adaptation indicators drawn from organizational survey instruments calibrated to the specific shifts that AI transformation requires, and coordination efficiency measures that track the resolution of the coordination cost inflation that the transition period produces; these leading indicators provide the early warning system that allows leadership to distinguish a healthy J-curve from a failing transformation program, and to make the targeted interventions that keep the transformation on track before the financial evidence of a problem has compounded to the point where intervention is more expensive and organizationally disruptive.

The *second implication* is that the investment protection mechanisms governing AI transformation programs must be designed to resist the short-term financial pressure that the J-curve's downward stroke reliably generates, because the investments most vulnerable to budget pressure during the trough period (e.g., the process redesign investment, the capability development investment, the cultural adaptation investment) are precisely the investments whose reduction most severely damages the probability of reaching the inflection point. Building explicit investment protection mechanisms into the program governance framework such as board-level commitments to maintain intangible capital investment through defined performance thresholds, CFO-level financial frameworks that distinguish intangible capital formation from operational cost, and CEO-level communication that prepares the organization for the J-curve's shape before it is entered rather than explaining it after it has arrived are not financial inflexibility. It is the governance architecture that makes financial discipline and strategic commitment simultaneously achievable.

The *third implication* is that the timeframe over which AI transformation ROI is evaluated must be calibrated to the J-curve's actual return profile rather than to the quarterly and annual reporting cycles that govern most financial performance assessment. Why? Because an AI transformation program evaluated against quarterly financial targets during the trough period will always appear to be failing, regardless of the quality of the intangible capital it is building, and the leadership that makes investment decisions based on that appearance will consistently make the wrong decisions at the most consequential moments.

The organizations that will define the AI productivity frontier of the next decade are not those that avoid the J-curve (no organization that pursues genuine AI transformation can avoid it) but those that enter it with the clearest understanding of its shape, manage through it with the most disciplined combination of investment commitment and organizational development capability, and emerge from it with the most fully formed intangible capital base that the upward stroke can build upon. The J-curve is not a risk to be mitigated or a cost to be minimized but rather the price of genuine transformation, in other words the organizational tuition that separates the enterprises that use AI to incrementally improve existing ways of working from those that use AI to build genuinely new organizational capabilities that generate compounding competitive advantage over time. Paying that tuition willingly, wisely, and with clear eyes about what it buys is among the most important acts of strategic leadership that the AI era demands.

The Buy vs. Build vs. Tune Decision Framework: A Methodology for Choosing the Right AI Development Strategy

Among the most consequential and most frequently mishandled decisions in enterprise AI strategy is one that appears, on its surface, to be a straightforward procurement choice: should the organization purchase access to an existing AI capability through a vendor API or SaaS platform, fine-tune an existing open-source or commercially available foundation model on its own data and for its specific use case, or build a proprietary AI architecture from the ground up using its own data, its own engineering talent, and its own computational infrastructure? The decision matters enormously because each of these three paths carries fundamentally different implications for cost structure, time to value, competitive differentiation, data sovereignty, organizational capability development, and long-term strategic flexibility, all implications that interact with the organization's specific competitive position, data assets, regulatory environment, and AI maturity in ways that make the right answer genuinely different for different organizations facing apparently similar AI challenges.

What makes the decision particularly difficult is that it is rarely presented to leadership in its full strategic complexity. Instead the decision is dressed as a vendor selection question, a technical architecture discussion, or a procurement negotiation, managed primarily by technology and procurement functions that have the expertise to evaluate the technical and commercial dimensions of the alternatives but may lack the strategic framework to connect those dimensions to the competitive, economic, and organizational considerations that should be most decisive. The result, in a distressing proportion of enterprise AI programs, is a build-buy-tune decision made primarily on the basis of vendor relationship quality, technology team preferences, or the path of least organizational resistance, rather than on the basis of a rigorous analytical framework that connects the decision to the organization's specific strategic context. This section encourages that framework: a structured, sequential methodology for making the build-buy-tune decision with the analytical rigor that its strategic consequences demand.

The Strategic Triad: Navigating Build vs. Buy vs. Tune in Enterprise AI

Before the build-buy-tune decision can be made intelligently, it must be understood in terms of what each option in-fact involves. Why? Because the vocabulary of these three alternatives is used with sufficient looseness in enterprise AI discussions that the same words frequently describe significantly different technical and organizational realities. *Buying* AI capability (the API or SaaS option) means accessing AI functionality developed, trained, maintained, and continuously improved by an external vendor, typically through a subscription or consumption-based commercial relationship that provides access to the vendor's model without providing access to the model's weights, training data, or underlying architecture. This option encompasses a spectrum that ranges from narrow, task-specific AI APIs such as sentiment analysis services, optical character recognition, speech-to-text transcription to the broad, general-purpose large language model APIs offered by OpenAI, Anthropic, Google, and others that can be prompted to perform a wide range of cognitive tasks without any organization-specific customization.

Fine-tuning (the tune option) means taking an existing foundation model whose weights are accessible, either through open-source release or through a commercial fine-tuning program, and adapting it to the organization's specific domain, vocabulary, task requirements, or performance characteristics by training it further on organization-specific data. Fine-tuning does not require building a model from scratch (i.e., the foundation model's general capability is preserved and adapted rather than replaced) but it does require meaningful ML engineering capability, access to high-quality organization-specific training data, and the computational infrastructure and MLOps tooling to manage the fine-tuning process and the resulting customized model's lifecycle. *Building* (the proprietary architecture option) means developing AI capability from the foundational level: designing the model architecture, assembling and curating the training dataset, executing the training process, and deploying and maintaining the resulting model entirely within the organization's own infrastructure and under its own control. This option is the most expensive, the most technically demanding, and the slowest to reach production, but it is also the option that provides the highest degree of competitive differentiation, data sovereignty, and strategic control when the organizational and competitive conditions that justify it are genuinely present.

The analytical framework for choosing between these options is most usefully structured as a sequential decision process that moves through five analytical dimensions in a specific order, i.e., an order that reflects the logical priority of the considerations that should be most decisive, ensuring that the most strategically important factors are evaluated before the more technically specific ones that are easier to quantify but less determinative of the decision's long-term consequences. The typical five dimensions include: 1) competitive differentiation; 2) data sensitivity and sovereignty; 3) performance requirements and customization depth; 4) organizational capability and cost; and 5) strategic flexibility and vendor risk. Working through these dimensions in sequence produces a decision that is anchored in strategic logic rather than technical preference, and that can be defended to boards, CFOs, and business unit leaders in terms that connect the technical architecture choice to the business outcomes it is designed to produce.

The First Analytical Dimension

Competitive differentiation asks the foundational strategic question that should govern the entire decision: does the AI capability being developed represent a source of genuine and durable competitive advantage that justifies the investment required to own and control it, or is it a capability that the organization needs to have but that does not differentiate it from competitors in ways that customers value and that AI markets reward? This distinction is the most important and most frequently overlooked in build-buy-tune decisions, because the instinct to build proprietary AI capability (to own the models, control the architecture, and avoid vendor dependency) is organizationally compelling in ways that can override the strategic analysis that would reveal its cost to be unjustified. The principle that should govern this dimension is straightforward: AI capabilities that are industry-standard, functionally equivalent across vendors, and not meaningfully differentiated by the organization's specific data or domain expertise should almost always be bought rather than built or tuned, because the competitive advantage of building them is negligible and the cost is substantial.

A retail company that needs an AI system to generate product descriptions for its e-commerce catalog is not building a competitive moat by developing a proprietary language model for that task… the task is sufficiently generic that the leading commercial APIs perform it as well as any proprietary model the retailer could build, and the investment in building or even fine-tuning a proprietary model for this purpose would generate no competitive return that the API option could not provide at a fraction of the cost. Conversely, a financial services company that has developed a proprietary dataset of alternative credit signals from fifteen years of customer relationship data, signals that no competitor can replicate and no public model has been trained, is building a genuine competitive moat by developing AI underwriting capability that incorporates those signals in ways that a generic API cannot approximate, and the investment in fine-tuning or building proprietary models on that dataset is justified by the competitive differentiation it enables. The test that operationalizes this dimension is specific: if a competitor with equivalent budget could achieve equivalent AI capability by selecting the same vendor and signing the same contract, the capability is not a point of competitive differentiation and the buy option should be the strong default.

The Second Analytical Dimension

Data sensitivity and sovereignty addresses the regulatory, legal, and competitive risk considerations that determine whether submitting the organization's data to an external AI vendor's system is permissible, prudent, and consistent with the organization's data governance obligations. As the intellectual property and data sovereignty analysis in this text establishes in detail, the submission of sensitive, proprietary, or personally identifiable data to external AI APIs creates risks (e.g., data leakage into vendor training datasets, cross-border data transfer compliance obligations, intellectual property exposure in AI-generated outputs, and competitive intelligence vulnerability) that are not always adequately addressed by vendor data processing agreements and that can be decisive in the build-buy-tune decision regardless of where the competitive differentiation analysis points. Organizations operating under regulatory frameworks that impose strict data residency requirements (e.g., GDPR's restrictions on cross-border personal data transfers, HIPAA's requirements for protected health information handling, the EU AI Act's data governance obligations for high-risk AI systems) may find that the API option is simply not available for certain use cases, not because it is technically inferior but because the data processing architecture it requires cannot be made compliant with applicable regulatory requirements at acceptable cost and complexity.

Similarly, organizations whose competitive advantage is substantially based on proprietary data assets such as the customer behavior databases, the operational process data, the domain-specific knowledge repositories that represent the accumulated value of years of market presence also must assess carefully whether exposing those assets to external AI systems, even under contractual protections, represents an acceptable competitive risk. For these organizations, the tune or build options which keep the data within the organization's own infrastructure may be strategically necessary regardless of their higher cost, because the alternative is exposing the data moat that is the foundation of the organization's competitive position to a risk of leakage or misappropriation that the moat's competitive value makes unacceptable.

The Third Analytical Dimension

Performance requirements and customization depth evaluates whether the task-specific performance the organization requires from its AI system can be achieved through a general-purpose API, or whether it requires the domain-specific adaptation that fine-tuning provides, or the fully custom architecture that building enables. General-purpose foundation model APIs perform extraordinarily well across a broad range of common cognitive tasks like text generation, summarization, classification, question answering, code generation and the performance gap between a well-prompted general API and a fine-tuned model on these common tasks has narrowed substantially as foundation model capability has advanced. The performance gap remains significant, however, for tasks that involve highly specialized domain vocabulary, unusual task formats, strict output structure requirements, or performance standards that require the model to have internalized domain-specific knowledge that is not adequately represented in the general pre-training corpus.

A legal AI system that needs to identify specific clause types in complex commercial contracts with the precision that legal risk management requires will typically outperform a general API with a fine-tuned model trained on a curated corpus of similar contracts not because the general API lacks language understanding capability, but because the specialized vocabulary, the subtle distinctions between legally significant clause variations, and the domain-specific judgment required for high-confidence legal classification are not adequately developed through general pre-training and require the domain-specific learning that fine-tuning on representative legal documents provides. The performance dimension also encompasses latency, throughput, and reliability requirements that the API option may not be able to satisfy: applications that require sub-second inference at high transaction volumes, that must operate reliably in environments with limited or unreliable internet connectivity, or that require guaranteed availability guarantees that vendor SLAs do not provide, may need the on-premises or private cloud deployment that the tune or build options enable. Evaluating this dimension requires honest performance benchmarking (i.e., testing the API option against real organizational data and real task requirements before concluding that it is insufficient) because the natural human tendency to overestimate the performance improvement that custom development will provide and underestimate the performance that well-prompted general APIs can achieve is a consistent source of unjustified build decisions that deliver marginal performance improvement at substantial cost.

The Fourth Analytical Dimension

Organizational capability and total cost of ownership represent what primarily ground the decision in the organizational and financial realities that make the technically optimal option practically achievable or practically unachievable given the organization's specific situation. Fine-tuning and building require ML engineering talent, data engineering infrastructure, MLOps tooling, computational resources for training and serving, and the organizational governance processes for model lifecycle management; in sum a combination of capabilities that is expensive to build, difficult to attract and retain, and organizationally complex to manage.

The total cost of ownership calculation for the tune and build options must therefore include not just the direct costs of model development and deployment but the full cost of the organizational capability required to do those things well: the talent acquisition and retention costs for the ML engineers, data scientists, and ML platform engineers the approach requires, the infrastructure costs for the training and serving environment, the MLOps tooling costs, the ongoing model monitoring and retraining costs, and the governance and compliance costs of managing proprietary AI systems in production. For many organizations, particularly those in the early stages of AI maturity, those in industries where AI talent is particularly scarce and expensive, and those whose AI use cases are not sufficiently numerous or strategically significant to justify building the organizational infrastructure that proprietary model development requires, the total cost of ownership calculation will strongly favor the API option even for use cases where fine-tuning would provide meaningful performance improvement, because the organizational investment required to realize that performance improvement exceeds the business value it generates.

The organizational capability dimension also encompasses the risk of talent dependency or the strategic vulnerability that organizations create when their AI capability is concentrated in a small number of highly specialized individuals whose departure would materially impair the organization's ability to maintain or develop its AI systems. This risk is more acute for the build option than the tune option, and more acute for the tune option than the buy option and must be factored into the total cost of ownership calculation as an expected cost weighted by the probability and impact of key talent loss.

The Fifth Analytical Dimension

Strategic flexibility and vendor risk represent the mechanisms used to evaluate the long-term strategic implications of the dependency relationships that each option creates and the optionality it preserves or forecloses. The API option creates vendor dependency of a specific and potentially consequential kind: the organization's AI capability is contingent on the continued availability, pricing, performance, and strategic direction of the vendor whose API it uses, and changes in any of these dimensions (e.g., vendor pricing increases, API deprecation, capability degradation, vendor acquisition, or regulatory action against the vendor) can impair the organization's AI capability in ways that are difficult and expensive to address quickly. This vendor dependency risk is real and must be factored into the build-buy-tune decision, but it must also be assessed proportionately: for commodity AI tasks where multiple vendors offer equivalent API capability, the switching cost of moving from one vendor to another is manageable and the dependency risk is low.

For tasks where a single vendor's API provides substantially superior performance and that superiority is embedded deeply in the organization's products, processes, and customer experiences, the dependency risk is substantially higher and the mitigation cost may be worth incurring either through multi-vendor architecture or through the development of fine-tuned alternatives that reduce API dependency. The build and tune options create a different kind of strategic flexibility risk: the risk of capability lock-in to a specific model architecture or fine-tuning approach that becomes technically obsolete as the AI field advances, requiring costly migration to new architectures to remain competitive. Managing this risk requires an architectural discipline that many organizations building proprietary AI capability do not apply consistently, e.g., abstracting the AI capability layer from the application layer in ways that allow model components to be replaced without rebuilding the applications that depend on them, and maintaining the MLOps infrastructure and organizational capability required to retrain and redeploy models as the technical landscape evolves.

Synthesizing these five analytical dimensions into a decision requires a structured judgment process that weights the dimensions appropriately for the organization's specific context and resolves the tensions between dimensions that frequently point in different directions. The framework that emerges from this synthesis is not a simple decision tree because the interactions between dimensions are too complex and too context-dependent for a mechanical rule to capture them adequately… but it does produce a set of strong directional signals that should govern the decision in the majority of enterprise AI use cases. *Buy* is the strong default for AI capabilities that are not sources of competitive differentiation, that do not involve sensitive data that cannot be safely processed externally, that general-purpose APIs perform at acceptable quality levels, and that the organization does not have the capability or the strategic rationale to develop proprietary alternatives.

Tune is the appropriate choice for AI capabilities that require domain-specific performance improvement beyond what general APIs can provide, that involve data that can be processed in fine-tuning infrastructure with appropriate sovereignty controls, that represent genuine competitive differentiation when customized to the organization's specific domain, and that the organization has or can cost-effectively develop the ML engineering capability to execute and maintain. *Build* is justified only for AI capabilities that represent the core of the organization's competitive strategy, that require full architectural control to deliver the performance and data sovereignty that the competitive and regulatory context demands, and that the organization has the talent, infrastructure, and financial resources to execute at the quality level that production deployment requires. The most important discipline the framework imposes is the requirement that the build option (the most expensive, the most talent-intensive, and the most organizationally demanding of the three) be chosen only when the competitive differentiation analysis provides a genuinely compelling justification, and that the strong human preference for building over buying that technology teams and organizationally ambitious leaders consistently exhibit be treated as a source of decision bias to be corrected by rigorous analysis rather than a strategic instinct upon which to act.

Dynamic Governance and the Intangible Economics of AI Maturity

The build-buy-tune decision is not a one-time choice that is made at program inception and remains fixed throughout the AI program's lifecycle. It is a dynamic decision that should be revisited as the organization's AI maturity develops, as the vendor landscape evolves, as the competitive dynamics of the organization's industry shift, and as the specific AI capability in question matures from an early-stage competitive differentiator to a commodity that the market provides efficiently. The organizations that manage this decision most effectively treat it as an ongoing element of their AI portfolio governance — reviewing the build-buy-tune rationale for each major AI capability annually, assessing whether the competitive differentiation, data sovereignty, performance, capability, and strategic flexibility considerations that drove the original decision continue to support it, and making the transitions between options (from build to tune as open-source models improve, from tune to buy as APIs develop sufficient domain capability, from buy to build as a capability becomes sufficiently central to competitive strategy to justify proprietary development) that the evolving landscape makes appropriate. This portfolio-level dynamism in the build-buy-tune decision is one of the most reliable markers of AI strategic maturity, and its presence or absence in an organization's AI governance framework is a strong predictor of whether the organization's AI investment will generate the compounding strategic returns that sophisticated AI portfolio management enables or the static, suboptimal outcomes that decisions made once and never revisited reliably produce.

The practical implication for AI investment planning and ROI modeling is both clear and uncomfortable for organizations accustomed to technology investment frameworks built around hard cost and revenue metrics. A complete and honest accounting of AI return on investment must treat training, workforce development, process redesign, and organizational learning infrastructure not as overhead costs to be minimized in service of a cleaner ROI calculation, but as primary value-generating investments whose returns, while longer-dated and harder to measure, are in many cases larger and more durable than the direct automation or efficiency returns that dominate the financial model. This requires expanding the valuation framework to include metrics that capture intangible capital accumulation… AI literacy indices across the workforce, process redesign completion and quality measures, learning loop velocity metrics that track how quickly the organization incorporates AI deployment feedback into improved practice, and talent capability assessments that quantify the premium the organization's AI-augmented workforce commands relative to industry peers.

It also requires governance structures that protect intangible capital investments from the budget pressure that invariably mounts when AI programs encounter the value dip that precedes mature returns… the period during which the costs of transformation are fully realized but the compounding returns on intangible capital have not yet materialized in the financial statements. Organizations that cut training budgets, defer process redesign, and reduce change management investment in response to short-term ROI pressure at precisely this moment are not reducing cost. They are liquidating the assets that would have generated the returns they are now failing to achieve. The *economics of AI* are fundamentally the *economics of intangible capital formation*, and the organizations that understand this earliest will build the durable advantage that later entrants will find increasingly difficult and expensive to replicate.

The financial impact of AI on operating performance is significant, influencing both cost structures and revenue generation. Many AI applications directly reduce operating expenses through automation, intelligent process optimization, and improved resource allocation. Predictive analytics can streamline inventory management, minimize stockouts, and reduce carrying costs, thereby improving gross margins. Automated document processing and AI-enabled customer service can reduce labor intensity, accelerate cycle times, and lower overhead costs, producing tangible Earnings Before Interest, Taxes, Depreciation, and Amortization (EBITDA) uplift. EBITDA refers to a financial metric used to evaluate a company's *operational performance* by focusing on earnings generated from core business activities… before accounting for financing decisions, tax environments, and non-cash expenses. These efficiency gains translate into improved enterprise value, reinforcing the rationale for continued investment and integration.

Value Redefinition: Adaptive Budgeting and Holistic ROI in the AI Era

Beyond cost efficiencies, AI opens avenues for new revenue streams and business models that traditional technologies could not support. Personalization engines can drive higher conversion rates and customer lifetime value by tailoring offerings to individual preferences and behaviors. Dynamic pricing systems allow companies to adjust prices in real time based on demand fluctuations, competitive positioning, and market signals, unlocking revenue optimization at scale. In industrial sectors, outcome-based business models (such as predictive maintenance agreements tied to equipment uptime) enable novel forms of monetization that align provider incentives with customer outcomes. Incorporating these revenue innovations into financial models reveals AI's potential not only to enhance existing operations but to redefine value creation altogether.

Given the iterative nature of AI development, budgeting for AI initiatives requires a departure from conventional IT governance frameworks. AI projects often progress through cycles of experimentation, validation, and refinement, with insights from early phases shaping subsequent investment decisions. Rigid, fixed-budget approaches can impede innovation, while overly flexible spending can lead to waste. Organizations increasingly adopt phased funding, milestone-driven investment, and portfolio-based governance to balance discipline with agility. Such mechanisms allow promising projects to receive additional resources as evidence accumulates, while underperforming initiatives can be curtailed before accruing excessive sunk costs. This adaptive budgeting approach supports continuous learning while safeguarding financial prudence.

Measuring the return on AI investments demands an expanded view of value creation that encompasses both tangible and intangible benefits. Direct financial gains, such as cost savings, margin improvements, and incremental revenue, are essential indicators but do not fully capture AI's enterprise impact. Improved decision quality, enhanced customer experience, accelerated innovation cycles, and reduced operational risk contribute meaningfully to long-term competitiveness, even if they are less readily quantifiable. Multi-criteria evaluation frameworks and balanced scorecards allow organizations to integrate financial and non-financial metrics, providing a more comprehensive picture of AI performance. Such holistic assessments enable executives to understand the broader strategic implications of AI deployment rather than focusing narrowly on short-term returns.

Scenario planning plays a critical role in AI valuation due to the fluid and rapidly evolving technological landscape. Shifts in regulation, competitive dynamics, data availability, or model capabilities can rapidly alter the cost-benefit profile of an initiative. Developing best-case, base-case, and worst-case scenarios enables organizations to evaluate the resilience of investment decisions under varying conditions. Finance teams can use these scenarios to guide capital allocation, assess risk tolerance, and design contingency plans that preserve optionality even in adverse environments. The integration of scenario planning into AI economics ensures that strategic decisions remain robust in the face of technological uncertainty.

A comprehensive assessment of the economics of AI must also incorporate lifecycle costs. Unlike traditional software systems, AI models require ongoing maintenance, retraining, and performance monitoring to remain effective. Data drift, model degradation, and evolving business requirements *necessitate* continual updates. In addition, organizations must account for potential costs associated with ethical breaches, compliance failures, biased outcomes, or security vulnerabilities. These risks underscore the importance of evaluating *total cost of ownership* rather than focusing exclusively on initial deployment expenses. By considering long-term operational, regulatory, and governance obligations, organizations develop more accurate financial projections and ensure sustainable investment strategies.

Ultimately, integrating AI economics into corporate strategy requires collaboration between finance, business leadership, and technical teams. Chief Financial Officers and finance organizations must develop fluency in the unique characteristics of AI-related expenditures, the dynamics of usage-based pricing, and the principles of model lifecycle management. Likewise, AI and data science teams must understand the financial constraints and strategic imperatives that shape resource allocation. When financial rigor and technical innovation operate in concert, organizations can optimize investment decisions, accelerate value realization, and build enduring competitive advantage. Through this synthesis, AI becomes not only a technological asset but a central component of enterprise value creation and strategic differentiation.

CHAPTER 6

Competitive Strategy and AI Moats

"In the age of AI, the new oil is data, the new factory is the model, and the new castle is the closed loop that keeps both."
— Adapted from Andrew Ng and Satya Nadella, 2024–2025

Strategy in the Age of Intelligent Competition

COMPETITIVE STRATEGY HAS ALWAYS BEEN about creating advantages that are difficult for rivals to replicate, and also about sustaining those advantages as the competitive environment evolves. For most of the twentieth century, the primary sources of durable advantage were scale economies, brand equity, switching costs, and control of distribution channels. The classic frameworks of industrial organization economics and strategic management (e.g., Porter's Five Forces, the resource-based view, and transaction cost theory) were built to explain and exploit these structural sources of advantage.

Artificial intelligence is rewriting the rules of competitive strategy in ways that these classic frameworks partially capture but do not fully explain. Some of the most successful competitors today, namely Microsoft, Tesla, NVIDIA, JPMorgan Chase, Shopify, BYD, Eli Lilly, and some number of less visible but structurally dominant incumbents, did not simply deploy AI as a productivity tool. They redesigned their competitive architectures around AI, creating self-reinforcing advantage structures that have proven remarkably difficult for competitors to replicate.

This chapter develops a rigorous framework for understanding how AI creates and sustains competitive advantage (what practitioners have come to call ***AI moats***) and how organizations can build, defend, and extend those moats in an environment where AI capabilities are advancing rapidly and the threat of disruption is persistent. The analysis draws on strategic theory, empirical evidence from the current competitive landscape, and the emerging literature on AI-specific competitive dynamics.

The Anatomy of Competitive Advantage in the AI Era

Porter's Five Forces framework remains a useful starting point for competitive analysis: the threat of new entrants, the bargaining power of suppliers and buyers, the threat of substitute products, and rivalry among existing competitors together shape industry attractiveness and the distribution of value. The resource-based view complements this industry-level analysis by focusing on firm-level resources and capabilities that are valuable, rare, inimitable, and organizationally embedded, known as the VRIO framework. These frameworks do not become obsolete in the AI era, but they require significant extension to capture the distinctive dynamics of AI-based competition.

Several features of AI-based competition are inadequately captured by classical frameworks. *First*, AI creates feedback loops between usage, data accumulation, and capability improvement that have no analogue in traditional resource-based competition: the more a firm deploys its AI system, the better its AI system becomes, which enables more effective deployment, which generates more data….. a compounding dynamic that classical models do not explicitly represent. *Second*, AI shifts the locus of competitive advantage from physical assets and organizational processes to data assets, model architectures, and the engineering talent required to develop and maintain them, all forms of capital that have distinctive economic properties, including near-zero marginal cost and high transferability. *Third*, AI enables platform-based competition across industries that were previously structurally distinct, creating competitive dynamics that blur traditional industry boundaries.

Data Moats: The Foundation of Durable AI Advantage

As mentioned, the concept of an AI moat (adapted from Warren Buffett's moat framework for enduring competitive advantage) refers to a self-reinforcing structural advantage that makes it difficult for competitors to replicate a firm's AI-based capabilities. Like traditional economic moats, AI moats derive their durability from the difficulty competitors face in reproducing the assets, processes, and relationships that underlie the advantage. Unlike many traditional moats, however, AI moats have distinctive characteristics: they tend to strengthen over time as data accumulates and models improve; they often operate at multiple levels simultaneously (data, model, operational, and ecosystem); and they can be eroded not only by direct competition but by technological discontinuities that render existing data assets or model architectures obsolete.

Understanding the anatomy of AI moats requires distinguishing among their constituent elements. Data moats, model moats, operational integration moats, and ecosystem moats are analytically distinct since they have different economic properties, different development timelines, and different vulnerabilities. Organizations that understand these distinctions are better positioned to build moats deliberately, rather than allowing them to accumulate haphazardly, and to identify which elements of their moat are most at risk from competitive or technological change.

At the heart of AI moats lies proprietary data: the unique, high-volume, high-quality datasets that allow organizations to train AI models that competitors cannot easily replicate. The economic logic is compelling: AI model performance is in most domains a function of data quantity, data quality, and data diversity, and organizations that control large, exclusive, and continuously expanding datasets can train models that outperform those built on publicly available or synthetic data… advantages that persist and compound as additional data accumulates.

Not all data creates moats of equivalent strength. The strategic value of a dataset depends on several factors: its exclusivity (i.e., can competitors access equivalent data from other sources?); its volume and coverage (i.e., does it span the range of cases required for high-quality model training?); its quality and labeling (i.e., is it accurate, consistently formatted, and appropriately annotated?); its recency and update velocity (i.e., does it capture current patterns, or is it subject to rapid obsolescence?); and its behavioral richness (i.e., does it capture the full complexity of the phenomenon being modeled, including edge cases and distributional tails?). The most strategically valuable data assets combine high exclusivity with high volume, quality, and behavioral richness… characteristics that are most likely to be found in data generated by large-scale operational systems with direct customer interaction.

Data Flywheel Dynamics

The most powerful data moats are not static repositories but dynamic systems mentioned previously… what practitioners call data flywheels or virtuous cycles. A data flywheel operates as follows: more users or operational activity generates more data; more data enables better model training; better models improve product or service quality; improved quality attracts more users or operational activity; and the cycle reinforces itself. Organizations that have established data flywheels in their core domains enjoy compounding advantage: their moat grows stronger with each passing month, while competitors who have not established similar flywheels face an ever-expanding capability gap.

The retail recommendation engine provides a canonical illustration of this dynamic. A global e-commerce platform with hundreds of millions of active users generates billions of behavioral signals (searches, views, purchases, returns, and reviews) every day. These signals are used to train recommendation models that become progressively more accurate as the signal volume increases. More accurate recommendations drive higher conversion rates and greater customer satisfaction, attracting more users and increasing purchase frequency. The additional users and purchases generate more data, completing the flywheel cycle. A new entrant attempting to replicate this capability faces not only the challenge of matching the incumbent's current model quality but also the compound data deficit that accumulates while the incumbent's flywheel continues to spin.

Despite their power, data moats are subject to several important vulnerabilities that strategic planners must explicitly address. Data moats can be eroded by changes in the phenomena being modeled: if consumer behavior, market conditions, or operational processes change sufficiently rapidly, historical data may become unrepresentative of current patterns, reducing the value of accumulated data assets. This vulnerability is particularly acute in domains subject to rapid structural change… such as financial markets, consumer preferences, or regulatory environments.

Data moats can also be circumvented by foundation models and transfer learning techniques that allow organizations to build capable AI systems without large domain-specific training datasets. A new entrant that fine-tunes a powerful foundation model on a modest amount of proprietary data may be able to achieve competitive capability levels in some domains without needing to match the incumbent's data accumulation. The strategic implication is that data moats are more durable when they are combined with other moat elements (e.g., model moats, operational integration, and ecosystem leverage) than when they stand alone.

Finally, data moats can be challenged through regulatory intervention: data portability requirements, data access mandates, and restrictions on data collection and use are features of the evolving regulatory landscape in the EU, US, and other jurisdictions. Organizations whose moats depend significantly on data exclusivity must develop strategies for managing regulatory risk, including compliance planning, engagement with regulatory processes, and contingency planning for scenarios in which current data practices become legally restricted.

Model Moats: Proprietary Intelligence as Competitive Differentiation

Proprietary AI models, when continuously trained on exclusive data, represent a second and distinct layer of competitive moat. The economic logic differs subtly from data moats: while data moats derive their strength from the exclusivity and volume of inputs, model moats derive their strength from the quality of the learning process applied to those inputs… the modeling architectures, training methodologies, evaluation frameworks, and domain expertise that allow an organization to extract superior intelligence from its data.

In domains where modeling expertise is scarce and the gap between state-of-the-art and off-the-shelf approaches is large, model moats can be significant. Autonomous driving provides a compelling illustration: the gap in performance between leading organizations and fast followers is not attributable solely to differences in training data, but also to differences in perception architectures, planning algorithms, simulation methodologies, and the accumulated engineering expertise required to operate at the frontier of a technically demanding domain. Similarly, in drug discovery, the difference between organizations that have developed deep expertise in AI-assisted molecular simulation, and those that have not, is not simply a matter of data access; it reflects *years* of methodological development and domain knowledge integration that cannot easily be replicated through off-the-shelf tools.

Model moats face a distinctive threat that data moats do not: the rapid commoditization of modeling capabilities through open-source releases, foundation model APIs, and the diffusion of machine learning expertise across the labor market. The history of deep learning since 2012 has been one of rapid capability diffusion: techniques that conferred significant competitive advantage in one year were widely accessible to practitioners at all levels within two to three years. The release of powerful open-source foundation models, including Meta's Large Language Model Meta AI (LLaMA), Mistral, and their derivatives, has dramatically lowered the barrier to deploying competitive language model capabilities.

The strategic implication is that model moats built primarily on architectural innovation or training technique are inherently time-limited and must be continuously renewed through sustained investment in research and development. Organizations that rely on model moats without reinforcing them with data moats, operational integration moats, or ecosystem moats are vulnerable to rapid competitive erosion. The most *defensible* model moats are those in which proprietary models are trained on exclusive data that cannot be replicated through foundation model fine-tuning, and in which the models are embedded in operational workflows that create switching costs reinforcing the competitive position.

While general-purpose AI capabilities are rapidly commoditizing, domain-specific model excellence remains a significant source of competitive differentiation in many contexts. Organizations that invest in building AI models precisely calibrated to the specific characteristics of their domain (including domain-specific data distributions, evaluation criteria, regulatory constraints, and operational requirements) can develop capabilities that generic models cannot match. In clinical diagnostics, financial risk modeling, materials science, and supply chain optimization, the difference between a domain-specialized model and a general-purpose model fine-tuned on limited domain data can be decisive, particularly in high-stakes applications where precision and reliability requirements are demanding.

Domain-specific model excellence is most durable when it is reinforced by data moats: organizations whose models are trained on proprietary domain-specific datasets that competitors cannot access combine the advantages of both moat types. The organizational capability to continuously update, evaluate, and improve domain-specific models (in response to new data, evolving requirements, and advances in modeling techniques) is itself a form of moat, since it requires a combination of technical expertise, domain knowledge, and organizational process that takes years to develop.

Operational Integration Moats: Embedding AI into the Core

Perhaps the most underappreciated dimension of AI moats is *operational integration*: the degree to which AI systems are embedded into an organization's core processes, decision workflows, and physical infrastructure in ways that create substantial switching costs. Organizations that have deeply integrated AI into their operations are not merely more efficient than competitors… they have restructured their operational architecture around AI in ways that would be prohibitively expensive and organizationally disruptive for competitors to replicate.

The strategic logic is analogous to the switching costs that arise from deeply embedded enterprise software systems, but more powerful in two respects. *First*, AI systems that are integrated into operational workflows generate usage data that continuously trains and improves the AI, thus creating a compounding advantage that static software systems do not possess. *Second*, AI integration often involves the redesign of human workflows, organizational structures, and decision rights in ways that go beyond technical implementation, creating organizational capital that is genuinely difficult to replicate. Competitors face not only the technical challenge of building equivalent AI systems but also the organizational challenge of restructuring their workflows, retraining their workforces, and redesigning their decision processes… a multi-year undertaking.

Vertical integration, meaning the control of multiple stages of the value chain, significantly amplifies the strength of AI moats by enabling end-to-end data capture, closed-loop optimization, and the elimination of the data access restrictions that constrain partially integrated competitors. In manufacturing, an AI-driven predictive maintenance system is most effective when the same organization designs the equipment, instruments it with sensors, operates it in production, monitors it through AI, and services it based on model recommendations, thus creating an unbroken data and feedback loop that produces progressively superior maintenance predictions.

Tesla's competitive architecture in the electric vehicle and autonomous driving domains illustrates this dynamic with particular clarity. By controlling vehicle design and manufacturing, software development, over-the-air update infrastructure, charging network operations, and insurance services, Tesla captures behavioral and operational data across the full customer lifecycle. This data feeds AI systems that improve vehicle performance, energy management, manufacturing quality, and autonomous driving capability… with each improvement reinforcing the data flywheel and creating capabilities that competitors who rely on third-party supply chains, software providers, or service partners cannot easily replicate.

The strategic value of *vertical integration* for AI moats depends on the specificity and exclusivity of the data generated at each stage of the value chain. Integration stages that generate uniquely informative data, such as sensor data from proprietary hardware, transaction data from direct customer relationships, or operational data from proprietary manufacturing processes, are most valuable. Integration stages that generate data equivalent to what competitors can access through third-party relationships add less incremental moat value, even if they create operational efficiencies.

A particularly powerful form of operational integration moat emerges at the intersection of physical and digital systems or what some analysts have called the *physical-digital interface*. Organizations that can instrument physical processes with sensors, actuators, and connectivity, and that can close the loop between AI-generated insights and physical operational responses, have access to categories of data and operational leverage that purely digital competitors cannot replicate. Industrial operators that instrument manufacturing lines with AI-enabled quality monitoring systems, agricultural enterprises that deploy AI-driven precision farming systems, and healthcare organizations that integrate AI with clinical devices and patient monitoring infrastructure are all building moats at the physical-digital interface intersection.

The strategic value of physical-digital integration moats derives from their combination of data exclusivity and operational embeddedness. The physical infrastructure that generates the data is often expensive and slow to replicate; the data itself is exclusive to the organization that owns the infrastructure; and the AI systems integrated with the infrastructure are embedded in operational processes that create substantial switching costs. Together, these characteristics create moats that are among the most durable in the AI landscape.

Ecosystem Moats: Network Effects and Platform Dominance

Platform dynamics increasingly define competitive landscapes in AI-intensive industries. Companies that develop AI-enabled platforms, those that connect developers, users, data providers, and complementary service providers in a shared infrastructure, can orchestrate ecosystems that generate network effects at multiple levels simultaneously. Understanding these platform dynamics, and the specific mechanisms through which they create and sustain competitive advantage, is essential for both incumbents seeking to defend platform positions and challengers seeking to disrupt them.

An AI platform *creates value* at multiple levels: for *end users*, who benefit from AI capabilities they could not develop independently; for *developers*, who can build applications and services on top of platform AI capabilities without replicating the foundational infrastructure; for *data providers*, who can contribute data in exchange for insights or revenue sharing; and for the *platform operator*, whose position at the center of the ecosystem creates data aggregation advantages, revenue diversification, and competitive information about emerging use cases. The more participants engage with the platform, the richer the data generated, the better the underlying AI models become, and the more valuable the platform is to all participants… a multi-sided network effect that compounds over time.

Network effects in AI ecosystems operate through several distinct mechanisms that strategic analysts must distinguish. Direct network effects arise when the value of the platform to each participant increases as the total number of participants increases; this is the classic network effect that characterizes social networks and communication platforms. In AI ecosystems, direct network effects are often present at the data level: more users generating more behavioral data improves model quality in ways that benefit all users.

Indirect or cross-side network effects arise when growth on one side of the platform creates value for participants on another side. In developer platform ecosystems, more developers building applications creates more value for end users, which attracts more end users, which attracts more developers… a reinforcing cross-side cycle. In data marketplace ecosystems, more data providers create more value for AI developers, which attracts more

AI developers, which creates more demand for data, which attracts more data providers. The strategic importance of these cross-side dynamics is that they create platform lock-in for all participant categories simultaneously, making the ecosystem progressively more difficult for competitors to displace.

Data network effects, a form of indirect network effect specific to AI platforms, arise when more users generate more data that improves AI model quality, which creates more value for users. These are among the most strategically significant effects in AI-based competition, because they create *compounding* advantage that accelerates as platform scale increases. Organizations building AI platforms should explicitly design for data network effects, ensuring that the system architecture captures and leverages usage data in ways that improve model quality rather than simply accumulating unused behavioral logs.

The *governance* of AI ecosystems is itself a source of competitive advantage and a significant source of competitive risk. Platform operators that establish favorable terms for developers, provide high-quality developer tools and documentation, offer transparent and stable API contracts, and create credible intellectual property protections for third-party applications can attract larger and more active developer ecosystems than those that treat third-party developers as extractable resources. The competitive histories of mobile operating system platforms, cloud computing platforms, and AI developer tools all illustrate the long-run costs of extractive ecosystem governance and the sustained advantages that accrue to platforms that invest in developer trust.

AI-specific governance challenges include the management of model access policies, the allocation of data rights between platform operators and ecosystem participants, the governance of responsible AI practices across the developer ecosystem, and the management of competitive tensions that arise when platform operators use ecosystem data to inform their own product development. The resolution of these tensions (through governance frameworks, contractual commitments, and institutional design) has significant implications for ecosystem health, developer trust, and the long-run durability of ecosystem-based competitive advantage.

Competitive Timing: First-Mover Advantage, Fast Following, and Strategic Windows

Strategic positioning in AI-intensive competition requires careful consideration of timing. First-mover advantage can confer substantial benefits when early entrants capture critical data assets, establish user bases that generate data flywheel effects, and define the technical or commercial standards that shape market development. The competitive histories of ride-sharing, food delivery, digital mapping, and recommendation systems all illustrate the durability of first-mover advantages that are grounded in data accumulation: organizations that entered these domains early accumulated behavioral datasets that late entrants could not replicate, even with superior technical capabilities.

The magnitude of first-mover advantage in AI competition varies significantly by domain and depends on several structural factors. Data scarcity, meaning the degree to which the relevant training data is generated by market participants rather than available from public or synthetic sources, amplifies first-mover advantage, since early entrants accumulate data that is unavailable to late entrants. Network effects amplify first-mover advantage by making established platforms progressively more attractive to additional participants. High switching costs, arising from operational integration or ecosystem embeddedness, extend the duration of first-mover advantage by reducing the rate at which established users defect to competing platforms. Regulatory first-mover advantages (e.g., early regulatory approvals in healthcare or financial services that competitors must separately obtain) can be among the most durable.

Note that being first to deploy AI in a domain is *not always* sufficient to confer durable competitive advantage, and the strategic literature on first-mover advantage cautions against conflating early entry with strategic leadership. Fast followers can capture significant market position by learning from the mistakes of early movers, leveraging more mature technology that has benefited from earlier development iterations, and deploying superior execution capability or regulatory navigation.

The conditions under which fast following is strategically viable depend on several factors. In domains where data accumulation effects are modest (because relevant training data is available from public or synthetic sources, or because the relevant phenomena are sufficiently stable that historical data does not lose value rapidly) "fast followers" with superior technical capability or operational execution can close the performance gap with early movers. In domains where regulatory approval is the primary barrier, fast followers who can navigate approval processes more efficiently may be able to enter after early movers have established clinical or operational proof-of-concept but before they have established dominant market share. In domains where foundation models substantially reduce the data requirements for competitive AI performance, fast followers may be able to achieve competitive capability levels on relatively modest domain-specific data.

One particularly important concept in AI competitive strategy is the strategic window: a period during which the structural conditions for moat construction are favorable, and after which the opportunity to build durable advantage in a specific domain may be significantly diminished. Strategic windows in AI are created by the convergence of technological maturity, data availability, regulatory clarity, and market receptivity that together make large-scale AI deployment feasible and commercially attractive.

Organizations that recognize and exploit strategic windows, such as committing resources to AI moat construction when the structural conditions are favorable, can establish positions that are progressively harder for later entrants to challenge. Organizations that hesitate, waiting for greater certainty or for competitors to validate the opportunity, may find that the window has closed: data flywheel effects have accelerated incumbents' advantage, ecosystem network effects have consolidated around established platforms, and the organizational capital required for deep operational integration has begun to accumulate among first movers. Developing the acute sensing capability to identify strategic windows early (and the organizational agility to commit resources rapidly when windows open) is a distinctive competence of firms that have demonstrated sustained advantage in AI competition.

Industry Disruption Patterns: AI Reshaping Competitive Landscapes

AI disruption follows distinctive patterns that vary across industries depending on the structure of data assets, the intensity of regulatory oversight, the degree of customer switching costs, and the availability of AI talent and infrastructure. Understanding these industry-specific patterns allows organizations to anticipate competitive threats, identify strategic opportunities, and develop positioning strategies calibrated to the specific disruption dynamics of their domain.

A useful analytical framework for AI disruption assessment evaluates industries across three dimensions: 1) *disruption attractiveness* (how large is the value at stake, and how tractable is the AI opportunity?), 2) *disruption vulnerability* (how susceptible is the incumbent competitive position to AI-based attack?), and 3) *disruption speed* (how rapidly are AI capabilities advancing relative to incumbents' ability to respond?). Industries that score high on all three dimensions (i.e., large value at stake, vulnerable incumbent positions, and rapidly advancing AI capabilities) face the most urgent disruption risk. Industries with lower scores on one or more dimensions provide more time for adaptive response, though complacency is rarely strategically justified.

Financial services experienced profound competitive disruption during the current period as AI-driven capabilities in credit underwriting, fraud detection, algorithmic trading, customer service automation, and regulatory compliance processing allowed digitally native firms to compete effectively against incumbents with superior balance sheets and established customer relationships. The structural vulnerability of traditional financial services incumbents arose from several sources: data assets that were technically proprietary but poorly integrated and difficult to exploit; legacy technology infrastructure that impeded rapid AI deployment; organizational cultures that were risk-averse in ways that slowed experimentation; and regulatory complexity that incumbents navigated through established relationships but that new entrants navigated through purpose-built compliance architecture.

The firms that emerged as competitive leaders in AI-intensive financial services competition, including JPMorgan Chase among incumbents and a range of digitally native fintech firms, shared a common strategic characteristic: they committed to treating AI as a core organizational capability rather than a technology deployment, investing in data infrastructure, talent, and organizational transformation at a scale commensurate with the competitive stakes. JPMorgan Chase's investment of billions of dollars in technology and AI infrastructure, combined with aggressive talent acquisition from technology firms, represents a strategic response to the AI disruption threat that few incumbent financial institutions have matched in scope or sustained commitment.

Likewise, the healthcare industry saw AI capabilities reshape competitive dynamics across multiple dimensions during the current period. In clinical applications, AI-assisted diagnostic imaging systems demonstrated performance competitive with specialist radiologists in specific domains, AI-driven pathology analysis accelerated cancer detection and characterization, and clinical decision support systems began to influence treatment protocols in leading health systems. In operational applications, predictive patient flow management, AI-driven revenue cycle optimization, and intelligent scheduling systems created significant efficiency advantages for early adopters.

The competitive dynamics of AI in healthcare are shaped by several distinctive structural features. Regulatory approval requirements for clinical AI systems (specifically, FDA clearance or approval in the US context) create both significant barriers to entry and durable competitive advantages for firms that have successfully navigated the approval process. Proprietary clinical datasets, those generated by high-volume healthcare systems with longitudinal patient records and diverse case mixes, create data moats that are particularly difficult to replicate. Moreover, the life-and-death stakes of clinical AI deployment create competitive dynamics in which trust, regulatory track record, and clinical validation evidence are as important as raw model performance.

Additionally, Eli Lilly's strategic deployment of AI in drug discovery and clinical development illustrates the competitive potential of AI-native approaches in pharmaceutical development. By integrating AI into molecular design, clinical trial optimization, and patient stratification, Lilly has accelerated development timelines and improved the predictive validity of clinical stage-gating decisions, thus creating competitive advantages in innovation productivity that accumulate over time. The pharmaceutical industry's long development cycles mean that the competitive impact of AI-driven R&D investments will compound for decades, making early strategic commitment particularly valuable.

Transportation and logistics AI disruption has progressed along two largely parallel tracks: the development of autonomous vehicle and robotics capabilities that could eventually restructure the economics of physical goods movement, and the nearer-term deployment of AI-driven optimization systems that are already delivering significant efficiency advantages across routing, demand forecasting, inventory positioning, and fleet management. The latter category has produced concrete and measurable competitive differentiation in the current period, while the former remains subject to significant technical, regulatory, and commercial uncertainty.

BYD's emergence as a globally competitive electric vehicle manufacturer illustrates the interaction between AI capabilities and broader industrial strategy. By vertically integrating battery technology, vehicle manufacturing, software systems, and charging infrastructure, BYD has accumulated proprietary data across the full EV ecosystem… enabling AI-driven improvements in battery management, range optimization, charging network management, and manufacturing quality that reinforce its competitive position. The combination of vertical integration, data flywheel dynamics, and government-supported scale economics has created a competitive position that Western incumbents have found difficult to challenge.

Retail AI disruption has been driven primarily by the application of AI to demand forecasting, inventory optimization, dynamic pricing, personalized recommendation, and supply chain management. The competitive divide between AI-enabled and AI-lagging retailers has widened significantly during the current period, with AI-capable operators demonstrating measurably lower inventory carrying costs, higher promotional effectiveness, and superior customer retention.

Shopify's evolution from an e-commerce platform to an AI-infused commerce operating system illustrates how platform strategies combined with AI can redefine competitive positioning in retail. By embedding AI capabilities (including AI-driven inventory management, fraud detection, product recommendations, and merchant analytics) into the core platform, Shopify has created value for its merchant ecosystem that is difficult for standalone retailers or smaller platforms to match. The data generated by millions of merchant transactions creates a flywheel that continuously improves Shopify's AI capabilities, while the improving capabilities attract more merchants, generating more data… a textbook data network effect operating at platform scale.

Pricing Power, Dynamic Pricing, and AI-Enhanced Market Intelligence

The accumulation of unique behavioral and transactional data strengthens pricing power by enabling organizations to understand customer willingness-to-pay with precision that less-informed competitors cannot match. Firms with richer data can segment customers more accurately, design pricing structures that capture a larger share of consumer surplus and dynamically adjust prices in response to competitive and demand signals at a speed and granularity that manual pricing processes cannot achieve. The competitive advantage created by AI-enhanced pricing capability is not limited to revenue optimization: superior pricing intelligence also enables more accurate demand forecasting, more efficient inventory management, and more targeted promotional investment.

Dynamic pricing in travel and e-commerce illustrates these dynamics with essential clarity. Revenue management systems in airlines, hotels, and ride-sharing platforms adjust prices in real time based on inventory levels, demand elasticity estimates, competitive pricing, and customer behavioral signals. The most sophisticated systems incorporate predictive models that anticipate demand shifts before they materialize, allowing preemptive pricing adjustments that capture revenue that reactive systems would forfeit. The more granular, high-frequency, and behaviorally rich the data inputs, the more refined the pricing strategy and the larger the revenue advantage relative to competitors with inferior data or less sophisticated optimization systems.

AI-enhanced pricing capabilities create competitive moats that are difficult for less data-rich competitors to challenge, but they also create regulatory and reputational risks that strategic planners must explicitly address. Algorithmic pricing systems that produce outcomes perceived as discriminatory, exploitative, or anti-competitive (even when the intent is pure revenue optimization) can attract regulatory scrutiny, antitrust investigation, and consumer backlash that erode both the pricing advantage and the broader competitive position. The regulatory landscape around algorithmic pricing is evolving rapidly, with competition authorities in the EU, US, and other jurisdictions developing frameworks for identifying and remedying algorithmic pricing coordination and discriminatory pricing practices.

Organizations that develop sophisticated AI pricing capabilities must therefore invest simultaneously in governance frameworks that ensure pricing systems operate within legal and ethical boundaries, and in regulatory engagement strategies that shape the development of pricing regulation in ways that reflect the legitimate benefits of AI-driven pricing efficiency. Pricing governance is not merely a compliance obligation but a strategic investment in the durability of pricing-based competitive advantage.

Defending Against AI Disruption: Incumbent Response Strategies

Established organizations confronting AI disruption face a variant of the classic innovator's dilemma: the investments and organizational capabilities required to build AI moats may be in tension with the structures and incentives that sustain current performance. Large incumbents often have significant advantages in data assets, customer relationships, brand equity, and regulatory experience… advantages that are potentially valuable inputs to AI moat construction. But they also face organizational inertia, legacy technology debt, talent structures poorly calibrated to AI development requirements, and financial incentives oriented toward near-term performance that can retard the long-term investment required for AI transformation.

The most common failure mode among incumbents confronting AI disruption is underestimating the pace and scope of competitive threat. Traditional competitive analysis tends to evaluate AI-native entrants against current-state performance metrics, missing the compounding dynamic of data flywheels and ecosystem network effects that makes the competitive threat grow rapidly over time. By the time AI-native competitors have accumulated sufficient scale to be visibly threatening by conventional metrics, the structural advantages they have built (e.g., data assets, model quality, operational integration, and ecosystem position) may already be difficult to overcome.

Incumbents confronting AI disruption have three primary strategic pathways for building AI capabilities: *internal development*, *strategic partnerships*, and *acquisition*. Each pathway has distinctive advantages and limitations in the context of moat construction, and the appropriate choice depends on the urgency of competitive threat, the availability of acquisition targets, the organization's existing technical capabilities, and the degree to which AI-based competitive advantage depends on proprietary data assets that cannot be acquired externally.

Internal development preserves data exclusivity and organizational control but requires significant time to build the talent, infrastructure, and organizational capabilities needed for competitive AI deployment. Strategic partnerships with AI platform providers, specialized AI firms, or research institutions can accelerate capability development but may compromise data exclusivity and create strategic dependencies that limit future flexibility. Acquisitions of AI-native firms can rapidly import talent, technology, and data assets but often face integration challenges that dilute the value of acquired capabilities and create cultural friction that impedes the organizational transformation required for AI-native competition.

The most effective incumbent response strategies typically combine all three pathways in a coordinated portfolio: building foundational AI infrastructure and talent internally; partnering for capabilities that are not strategically differentiating or where speed-to-deployment is paramount; and acquiring selectively for unique data assets, technical capabilities, or talent that would be prohibitively slow to develop internally. The coherent orchestration of these pathways, guided by a clear theory of which AI capabilities are core to competitive differentiation and which are commodity inputs, is a distinctive competence of incumbents that have navigated AI disruption successfully.

Ultimately, the most durable defense against AI disruption is *organizational transformation*: the systematic restructuring of talent, processes, culture, and governance around AI-native capabilities. Incumbents that treat AI as a tool to be applied incrementally to existing processes, while preserving the organizational structures, decision rights, and cultural norms of the pre-AI era, tend to fall progressively further behind AI-native competitors regardless of the specific AI tools they deploy. Incumbents that use AI disruption as a catalyst for broader organizational transformation (redefining roles, restructuring decision processes, redesigning workflows, and cultivating data-driven cultures) can leverage their existing assets in ways that AI-native startups cannot match.

Controlling the AI Stack: Infrastructure-Level Competitive Advantage

One of the most instructive competitive case studies of the current period is NVIDIA's emergence as the dominant supplier of AI computing infrastructure. NVIDIA's competitive position illustrates a form of AI moat that is distinct from the data, model, operational, and ecosystem moats discussed in earlier sections: infrastructure-level architectural control, in which dominance of the foundational computing layer creates a privileged position across the entire AI value chain.

NVIDIA's moat is built on the combination of hardware architecture, software ecosystem, and developer community that together constitute the Compute Unified Device Architecture (CUDA) platform. The CUDA ecosystem (developed over nearly two decades and comprising thousands of optimized libraries, developer tools, and application frameworks) creates a software moat that is extremely difficult for competitors to circumvent even when they develop GPU hardware with equivalent computational performance. AI researchers and engineers who have built their workflows around CUDA-optimized frameworks face substantial switching costs in migrating to alternative computing platforms, even if alternative hardware offers comparable raw performance at lower cost.

The strategic lesson of NVIDIA's competitive position is that control of platform layers (including computing infrastructure, software ecosystems, and developer tools) can create AI moats that are upstream of and complementary to the application-level moats discussed elsewhere in this chapter. Organizations that identify and control bottleneck resources in the AI value chain (including specialized compute, critical datasets, or essential developer tools) can build platform-level moats that extend their competitive influence far beyond their direct competitive domain.

The AI value chain encompasses multiple layers: silicon and compute infrastructure; cloud and data center infrastructure; foundational AI models and APIs; AI development tools and platforms; domain-specific AI applications; and end-user products and services. Competitive advantage can be pursued at any layer, but the layers differ significantly in their moat characteristics, capital requirements, competitive dynamics, and growth trajectories.

Infrastructure layers like silicon and cloud are characterized by massive capital requirements, significant economies of scale, and winner-take-most dynamics driven by the cost advantages of scale. Foundation model layers are characterized by intense research investment, rapid capability advancement, and significant uncertainty about the long-run distribution of commercial value between model developers and application builders. Application layers are characterized by domain-specific moat construction opportunities and closer proximity to end-user value, but also by competitive exposure to foundation model developers who may expand upstream into applications. Organizations seeking to identify their optimal positioning in the AI value chain must evaluate their existing capabilities, capital resources, and competitive context against the distinctive dynamics of each layer.

AI Moat Erosion: Threats, Vulnerabilities, and Sustaining Advantage

AI moats, like all competitive advantages, are subject to erosion over time. Understanding the forces that erode AI moats (and developing strategies to manage them) is as important as understanding how to build moats in the first place. The primary forces of moat erosion in AI-intensive competition are technological discontinuity, regulatory intervention, competitive imitation, and talent diffusion.

Technological discontinuity represents the most acute threat to existing AI moats. A sufficiently radical advance in modeling capability (such as the emergence of foundation models that dramatically reduced the data requirements for competitive natural language understanding) can devalue established data and model moats by enabling new entrants to achieve competitive performance without the accumulated assets of established players. Organizations with AI moats built primarily on one-dimensional advantages (i.e., data alone, or models alone, without operational integration or ecosystem embeddedness) are most vulnerable to technological discontinuity. Multi-layered moats, in which data, model, operational, and ecosystem advantages reinforce each other, are more resilient.

Competitive imitation accelerates as AI capabilities diffuse through the labor market, as open-source model releases reduce the technical barriers to entry, and as AI platform providers commoditize capabilities that were previously proprietary. Successful AI moat defense requires continuous investment in data acquisition, model improvement, operational deepening, and ecosystem development that keeps the moat reinforced against incremental competitive erosion. Organizations that treat AI moat construction as a one-time investment rather than a continuous competitive discipline tend to find their advantages steadily diminished.

Sustaining AI moats in a rapidly evolving technological landscape requires what can be called *adaptive moat architecture*: the organizational capability to continuously reassess the composition of one's competitive advantage, anticipate the evolution of the threat landscape, and proactively invest in moat elements that will remain defensible as technology and competition evolve. This is a fundamentally different organizational capability from the static moat optimization that might suffice in more slowly evolving competitive environments.

Organizations that have demonstrated sustained AI competitive advantage (including Microsoft in enterprise AI platforms, Google/DeepMind in research-to-product AI, and Amazon in AI-enhanced logistics) share a common characteristic: they invest continuously in exploring the next generation of AI capability while simultaneously deepening and defending their current competitive positions. They treat each successive wave of AI technology, from classical machine learning to deep learning to foundation models to agentic AI systems, as an opportunity to both extend existing moats and build new ones, rather than as a threat to be defended against.

One dimension of AI moat sustainability that is insufficiently appreciated in competitive strategy discussions is *governance*: the frameworks, processes, and institutional structures that determine how AI systems are developed, deployed, monitored, and adjusted over time. Organizations whose AI moats are built on capabilities that cannot be credibly governed (e.g., proprietary models whose decision logic is opaque to regulators, or data assets assembled in ways that may violate emerging privacy regulations) face regulatory risks that could rapidly and substantially erode their competitive positions.

Responsible AI governance is not simply a compliance burden but a strategic investment in moat durability. Organizations that invest in the transparency, fairness, and accountability of their AI systems build regulatory relationships and public trust that provide durable competitive advantages in a regulatory environment that is becoming progressively more demanding. Organizations that treat governance as an afterthought face the risk that regulatory intervention will dismantle moats that took years to construct... a competitive setback that cannot be quickly remedied.

The Ethics of AI Competitive Strategy

The pursuit of AI competitive advantage raises ethical questions that executives and strategists must confront directly, not merely as compliance matters but as genuine strategic and moral considerations. Several moat-building strategies that are economically attractive in the near term may create harms to consumers, competitors, employees, or society more broadly; business organizations have ethical obligations to consider and manage such harms apart from regulatory compliance requirements.

Data moats built on aggressive data collection practices (e.g., collecting personal data beyond what users reasonably expect, using behavioral data in ways that users do not understand or consent to, or acquiring data assets that were assembled in ways that violated individual privacy rights) may be technically legal in some jurisdictions while raising serious ethical concerns about the treatment of individuals as data sources rather than as autonomous agents with rights over their own information. The increasing regulatory attention to data collection and use practices reflects a broader social reckoning with these ethical concerns, and organizations that align their data strategies with the ethical expectations that regulation is moving toward will be better positioned for long-run competitive sustainability than those that maximize near-term data collection without regard to ethical constraints.

Some AI moat construction strategies raise antitrust and competition policy concerns that go beyond individual ethics to questions of market structure and competitive fairness. Business organizations that leverage AI moats to engage in predatory pricing, to exclude competitors from access to essential inputs, or to orchestrate pricing coordination through algorithmic systems may be building competitive positions that competition authorities will challenge as anti-competitive, regardless of their technical legality under current law. The regulatory scrutiny of AI-enabled market power is intensifying globally, and strategies that rely on competitive positions that may not survive regulatory examination are strategically fragile.

Executives and strategists have both ethical and strategic reasons to ask whether their AI moat-building activities are pro-competitive (e.g., driving innovation, improving consumer welfare, and expanding market opportunity) or anti-competitive, in the sense of foreclosing competition in ways that harm consumers and undermine market dynamism. The distinction matters not only for regulatory risk management but for the organizational culture and values that determine the long-run strategic character of the enterprise.

Practical Frameworks for AI Competitive Strategy

Strategic leaders need practical diagnostic tools for assessing the strength and composition of their organization's AI moat and identifying the dimensions that most require investment or defense. The following framework provides a structured approach to AI moat assessment across five dimensions: *data assets*, *model capabilities*, *operational integration*, *ecosystem position*, and *governance strength*.

For each dimension, the assessment should evaluate current position on a scale from *nascent* (limited or no moat element present) through *developing* (moat element present but not yet self-reinforcing) to *established* (self-reinforcing moat with meaningful competitive advantage) to *leading* (industry-defining position that shapes competitive dynamics for all participants). The assessment should also evaluate the trajectory of each dimension (e.g., is the moat strengthening, stable, or eroding?) and the primary threats to each dimension from competitive, technological, and regulatory forces.

Translating a moat assessment into an actionable strategic roadmap requires sequencing investments across moat dimensions in ways that reflect the organization's current position, resource constraints, and competitive timeline. Organizations at *early stages* of AI moat construction typically benefit from concentrating initial investments on data infrastructure and data flywheel development, since data moats take the longest to develop and provide the foundation for model and operational moats. Organizations with *established* data moats should prioritize model quality investment and operational integration deepening, which convert data assets into operational capabilities that are harder for competitors to replicate. Organizations with *strong and mature* data, model, and operational moats should prioritize ecosystem development and platform investment, which extend the organization's competitive influence beyond its direct operational domain.

Given the rapid pace of AI capability advancement and the uncertainty of competitive and regulatory evolution, scenario planning is a particularly valuable strategic discipline for organizations managing AI competitive positions. Effective AI scenario planning should explore at minimum three categories of uncertainty: technological discontinuity scenarios (what if a new modeling approach substantially reduces the value of our data assets or enables new entrants to achieve competitive performance without our accumulated advantages?); competitive response scenarios (what if our primary competitor acquires a transformative AI capability, executes a platform expansion into our core domain, or forms a strategic alliance that closes the capability gap with our position?); and regulatory change scenarios (what if data collection or use regulations substantially constrain our ability to expand or maintain our data moat?).

Business organizations that systematically explore these scenarios and develop strategic responses that preserve competitive relevance across a range of futures will be better positioned to make timely decisions when the competitive landscape evolves than those that rely exclusively on the continuation of current trends. The pace of AI development makes scenario planning not only a useful strategic exercise but an organizational necessity.

The Compounding Logic of AI Competitive Advantage

Competitive strategy in the AI era is characterized by a compounding logic that has no precise precedent in traditional strategic frameworks. Data flywheels accelerate data accumulation. Better data enables better models. Better models improve products and attract more users. More users generate more data. Each iteration strengthens the moat. Organizations that establish this compounding cycle early and invest continuously in each element of their moat architecture will accumulate competitive advantages that grow more formidable with each passing year.

The business organizations that will define competitive dynamics in AI-intensive industries over the next decade are not necessarily those with the largest budgets or the most technically sophisticated AI research programs. They are those that most clearly understand the specific moat architecture appropriate to their competitive context; that make the strategic commitments required to build that architecture before strategic windows close; that govern their AI capabilities with the transparency and accountability that long-run regulatory and social sustainability demands; and that develop the organizational capabilities (e.g., strategic sensing, adaptive planning, continuous investment discipline) required to sustain advantage as technology and competition evolve.

AI moats are not static fortifications but dynamic competitive systems that require continuous investment, adaptive management, and strategic renewal. Enterprise leaders who understand this, and who build the organizational disciplines required to sustain advantage in a rapidly evolving competitive landscape, will find that the compounding logic of AI competitive advantage creates durable and expanding positions that reward long-run strategic commitment.

The AI Moat Diagnostic Checklist

Use this five-question diagnostic to assess the current strength of an AI moat. Each question probes a distinct and essential moat dimension. Affirmative answers that reflect genuine structural advantage (not just aspirational capability) indicate a defensible competitive position. Organizations that cannot answer yes to at least two questions with genuine conviction do not yet have a moat and are at meaningful risk in AI-intensive competition.

- ☐ *Data Exclusivity*: Do we generate unique, high-volume behavioral or operational data that competitors cannot replicate through alternative sources, synthetic data, or foundation model pre-training?
- ☐ *Closed-Loop Flywheel*: Are we capturing every customer interaction, transaction, and operational signal and using it to materially improve our AI models on a continuous basis… such that our system is measurably better next month than it is today?
- ☐ *Deep Operational Embedding*: Have we integrated our AI systems so deeply into daily workflows, decision processes, and operational infrastructure that replacing them would be organizationally catastrophic and financially prohibitive for our customers or our own operations?
- ☐ *Physical-Digital Vertical Integration*: Do we control at least one mission-critical interface between physical operations and digital intelligence (e.g., manufacturing, logistics, clinical care, or energy management) where vertical integration generates exclusive data and operational leverage that competitors without that integration cannot access?
- ☐ *Platform or Ecosystem Control*: Do we operate a platform layer (e.g., data infrastructure, distribution, or developer tools) that external parties are actively building on top of, generating cross-side network effects and ecosystem lock-in that strengthen our position as the ecosystem grows?

Business organizations that answer "yes" to three or more of these questions with genuine structural conviction are positioned to compound competitive advantage for the remainder of this decade. The organizations that cannot are at material risk of becoming feature suppliers for someone else's AI platform… technically capable but strategically dependent.

CHAPTER 7

Data and Infrastructure Foundations

Data Infrastructure and AI Readiness: Foundations for Scalable Enterprise AI

EARLIER CHAPTERS INTRODUCED THE STRATEGIC imperatives behind enterprise AI adoption and examined the technological pillars that make modern AI possible. Building on those foundations, this chapter turns to the underlying data and infrastructure systems that transform conceptual AI capabilities into operational reality. These foundations constitute a multilayered combination of architecture, governance, computational resources, and organizational practices. Together, they determine whether AI initiatives remain theoretical or evolve into scalable, reliable, and value-generating systems. More importantly, this transformation is not purely technical; it demands coordinated changes to processes, culture, and responsibility structures across the organization.

The evolution from traditional analytics systems to the modern data stack marks a fundamental shift in how organizations collect, manage, and utilize information. Earlier architectures were generally built around relational databases, nightly extraction, transformation, and loading (ETL) jobs, and predictable, structured data environments. These systems were optimized for stable enterprise reporting workloads but were fundamentally limited in their ability to support the continuous, high-volume, multimodal data flows required for contemporary AI applications.

In contrast, the modern data stack is characterized by modularity, elasticity, and a design philosophy centered on adaptability. It integrates real-time ingestion frameworks, distributed storage layers capable of scaling to petabytes, compute engines that perform high-throughput transformations, and specialized machine-learning components that support experimentation and deployment. This modular design allows organizations to combine diverse technologies, ranging from open-source frameworks to cloud-native managed services, into integrated pipelines. Data moves through these layers via event-driven mechanisms and APIs, enabling teams across data engineering, analytics, and machine learning to collaborate efficiently.

A defining feature of this new architecture is its ability to manage unstructured and semi-structured data alongside structured enterprise datasets. AI systems often depend on text corpora, images, logs, sensor output, audio recordings, and other complex modalities, all of which require flexible storage formats and scalable computation engines. As AI use cases proliferate, organizations frequently shift from batch-oriented ETL workflows to streaming-first paradigms that support continuous model updating, dynamic grounding for generative systems, and real-time decision-making. This architectural evolution forms the operational backbone of modern AI and sets the stage for the more nuanced challenges of data quality, governance, and model readiness.

Although infrastructure provides the scaffolding, high-quality along with responsibly managed data is the substance that ultimately determines the reliability of AI systems. Data quality (referring to the accuracy, completeness, timeliness, consistency, and representativeness of the data source) is a prerequisite for building trustworthy models. Even sophisticated algorithms cannot compensate for flawed underlying data, and organizations that neglect quality control often face biased predictions, degraded user experiences, or regulatory exposure.

Ensuring data quality requires comprehensive observability. Contemporary practices include continuous profiling to detect anomalies, schema validation to prevent inconsistencies, and monitoring systems that track lineage, freshness, and drift over time. Combined, these mechanisms function as a critical form of preventative maintenance, reducing the likelihood of model degradation and ensuring that AI systems remain aligned with real-world conditions.

Closely intertwined with quality is the broader field of data governance. Governance frameworks establish clear ownership structures, enforce access controls, and ensure compliance with legal and ethical requirements. As regulatory environments evolve, as in encompassing privacy legislation such as GDPR, CCPA, and HIPAA, as well as emerging AI-specific governance regimes, organizations must embed responsible data practices into the *full lifecycle* of AI development. Techniques such as differential privacy, federated learning, and homomorphic encryption are increasingly used to protect sensitive information while preserving analytical utility. Governance also requires that model behavior is transparent and auditable, creating traceability between training data, model decisions, and business outcomes.

Labeling presents another essential element of the data foundation, particularly for supervised learning and fine-tuning. High-quality annotation remains a human-centered process, albeit one enhanced by automation and model-assisted labeling techniques. Effective labeling involves careful design of annotation guidelines, consistent reviewer workflows, measurement of agreement among annotators, and systematic quality checks. Increasingly, organizations are adopting active learning approaches in which models identify which samples should be labeled next, based on uncertainty or expected value, thereby improving efficiency and reducing costs.

Synthetic data has become an important complement to real-world datasets. Generated through rule-based systems, physics-informed simulations, or advanced generative models, synthetic data broadens the range of scenarios available for training. One finds it particularly beneficial when real data is scarce, imbalanced, sensitive, or costly to obtain. When integrated carefully with authentic data, synthetic datasets improve model robustness, protect privacy, and enable experimentation without risking exposure of personal or proprietary information.

Reliable AI systems depend on the ability to move data from source to model in a consistent, reproducible, and well-documented manner. Data pipelines serve as the *connective tissue* that orchestrates this movement. Modern pipelines incorporate both batch and streaming workloads, dynamically transform raw inputs into structured, model-ready outputs, and embed validation steps that ensure data integrity. As pipelines have grown more complex, orchestration frameworks have emerged that allow teams to define tasks, dependencies, and failure recovery procedures in a transparent and maintainable way. These workflows ensure that data engineers and machine-learning practitioners can iterate rapidly while maintaining operational reliability.

Feature engineering remains one of the most impactful activities within the ML lifecycle. In practice, improvements to features often yield greater gains than changes to model architecture. Effective feature engineering ranges from simple transformations, such as calculating ratios or aggregations, to advanced representations, such as embeddings for text, images, and logs. The growing emphasis on "data-centric AI" highlights the centrality of this work, encouraging teams to refine datasets and features through iterative, scientific processes.

Feature stores emerged to formalize and systematize this feature lifecycle. They function as centralized repositories in which features are defined, documented, versioned, and reused across teams and projects. By providing consistent APIs for both training and inference, feature stores eliminate discrepancies between offline and online computation, reduce redundant engineering effort, and ensure that models behave consistently in production environments.

Previously noted, the rise of *retrieval-augmented generation* (RAG) and context-aware AI systems has brought vector databases to the forefront of modern AI infrastructure. Vector databases store embeddings (i.e., high-dimensional numerical representations that encode semantic meaning) and support similarity search at scale. Through specialized indexing and approximate nearest-neighbor algorithms, vector databases enable AI systems to retrieve relevant information with low latency, even across billions of records. This capability is transformative for applications such as RAG pipelines, personalized recommendation engines, anomaly detection systems, and enterprise search tools. As AI increasingly depends on contextual information retrieved at inference time, vector databases have become indispensable components of the contemporary AI stack.

As AI projects transition from experimentation to production, organizations must adopt operational frameworks that ensure reliability, reproducibility, and safety. MLOps *extends* the principles of DevOps to the machine-learning lifecycle, integrating data engineering, model development, deployment, governance, and monitoring into a continuous workflow. At the heart of MLOps is the model registry, which tracks versions of models and their associated artifacts, enabling teams to compare performance across iterations, manage promotion through development stages, and maintain traceability of changes.

Lifecycle management increasingly involves automated training and retraining pipelines that respond to data drift, environmental changes, or shifts in business context. Monitoring systems track both technical metrics, such as latency and throughput, and behavioral metrics, including accuracy, bias indicators, and user satisfaction. These systems provide the feedback loops necessary to maintain long-term alignment between models and their intended functions.

Model deployment involves specialized infrastructure optimized for inference workloads. Modern serving frameworks support multiple model formats, scale elastically in response to demand, and incorporate GPU scheduling and resource optimization strategies. Because inference often represents the most significant ongoing cost in an AI system, organizations rely heavily on techniques such as quantization, pruning, distillation, and caching to reduce computational overhead without compromising accuracy. Advanced deployment strategies, including canary releases and blue–green deployments, ensure that new models can be introduced safely and gradually. In addition, the growth of edge computing has extended model deployment beyond the cloud, enabling AI to run on mobile devices, IoT hardware, and on-premise systems where low latency or data sovereignty are essential.

Cloud platforms have become the *dominant* environment for building and deploying AI systems, providing integrated suites of services that span storage, compute, training, deployment, monitoring, security, and governance. Providers such as AWS, Azure, and Google Cloud offer managed machine-learning platforms (e.g., SageMaker, Azure ML, and

Vertex AI, respectively) that simplify the complexity of provisioning infrastructure, orchestrating workflows, and scaling experimentation. These environments take advantage of specialized hardware accelerators, automated scaling mechanisms, and managed services that reduce operational overhead and allow teams to focus on model quality and business outcomes.

Cloud ecosystems are increasingly complemented by AI-native platforms that address the unique requirements of generative and retrieval-based systems. These platforms include model-serving engines designed for large language models, agent orchestration frameworks, embedding generation services, and fine-tuning toolkits for domain adaptation. They also provide tools for integrating RAG pipelines, managing long-term memory systems, and coordinating complex multimodal workflows. The rapid expansion of these ecosystems signifies a shift toward more modular, interoperable, and developer-friendly AI infrastructure, enabling organizations to innovate rapidly without reconstructing foundational components from scratch.

Distributed and edge computing further expand the architectural landscape by enabling models to execute closer to the source of data generation. This is increasingly important for applications requiring ultra-low latency, offline functionality, or enhanced privacy. As a result, organizations now commonly employ hybrid architectures that combine cloud, edge, and on-premise compute resources, selecting the optimal deployment pattern for each use case.

Security has become a central concern in AI infrastructure, not only because AI systems handle sensitive data but also because they introduce new classes of vulnerabilities. Threat vectors now include data poisoning, model extraction, prompt injection, adversarial examples, and exploitation of residual information encoded in model weights. Protecting against these threats requires a multifaceted approach that includes encryption, robust identity management, secure sandboxing of inference APIs, and continuous monitoring for anomalous behavior.

Risk management also intersects with governance, as organizations must ensure that AI systems comply with internal policies and external regulations. This responsibility involves maintaining auditable workflows, documenting model lineage, establishing guardrails for high-risk applications, and implementing human oversight mechanisms. Together, these measures create an environment in which AI can be deployed safely and effectively.

The rapid expansion of AI workloads has brought energy consumption and environmental impact into sharp focus. Training large-scale models can require substantial electrical power, while inference workloads (especially when deployed globally) introduce continuous and growing operational costs. As AI becomes more pervasive, sustainability is no longer a secondary concern but an essential aspect of long-term infrastructure planning.

Efforts to improve energy efficiency span hardware, software, and data-center operations. Hardware accelerators such as GPUs, TPUs, and dedicated *neural processing units* (NPUs) are designed to deliver higher performance per watt. Software-level optimizations (including model compression, quantization, pruning, and distillation) reduce the computational footprint without significantly affecting model performance. Data centers increasingly rely on renewable energy, advanced cooling mechanisms, and intelligent scheduling systems that distribute workloads to minimize energy peaks. At the application level, architectures such as edge inference and caching-based retrieval reduce unnecessary computation. Collectively, these strategies underscore that sustainable AI is both a technical and strategic imperative.

When viewed together, the elements described in this chapter, namely data architectures, governance frameworks, pipelines, feature systems, vector retrieval, MLOps practices, cloud ecosystems, security measures, and sustainability strategies, form an interconnected *ecosystem* rather than a collection of isolated components. An organization's ability to deploy AI effectively depends on resolving the interdependencies among these layers. Reliable data enables trustworthy models; robust infrastructure ensures scalability and operational continuity; governance provides ethical and regulatory alignment; and sustainability guarantees that AI growth remains economically and environmentally viable. As AI systems continue to expand in complexity and impact, this integrated perspective will remain essential to building resilient, responsible, and enterprise-ready AI capabilities.

Data Pipelines and Converged Architecture

The architecture of enterprise data systems is now undergoing its most fundamental transformation since the introduction of the relational database. For three decades, organizations built their data infrastructure around a principle of deliberate separation: transactional systems that processed the moment-to-moment activity of the business were kept architecturally distinct from the analytical systems that stored, aggregated, and reported on that activity. Between them sat the data pipeline, essentially the machinery responsible for extracting data from operational sources, transforming it into formats suitable for analysis, and loading it into the data warehouse where it could be queried for business intelligence. This architecture served its era well. But the era is *changing*, and it is changing in ways that make the old model not only suboptimal but increasingly unsuitable for the demands that modern AI-driven enterprise operations place on data infrastructure.

In the following section we examine three interconnected dimensions of that transformation. *First*, we'll explore how the convergence of transactional, analytical, and unstructured data environments is fundamentally altering the role and design of data pipelines. *Second*, we'll analyze the accelerating shift from batch-oriented to streaming-oriented pipeline architectures, and what that shift means for how organizations think about data latency, freshness, and the real-time responsiveness required by their AI systems. *Third*, we'll examine why converged architectural platforms are proving more naturally accommodating to AI workloads than the siloed or single purpose architectures they are replacing, and what this means for organizations building AI-driven capabilities across their business operations.

The Impact of Converged Data Environments on Data Pipelines

The traditional enterprise data architecture was built around what practitioners called the OLTP/OLAP divide, meaning the separation between Online Transactional Processing systems, which handled live operational data, and Online Analytical Processing systems, which handled historical analysis. This separation was not arbitrary. It reflected genuine technical constraints: the workload characteristics of transactional processing (i.e., high concurrency, fast writes, strict consistency) were fundamentally at odds with those of analytical processing, namely complex reads across large historical datasets, aggregations, and multi-table joins. Building a system that performed both workload types efficiently on the same infrastructure was, for most of the history of enterprise computing, not technically feasible at the scale enterprises required.

The data pipeline emerged as the engineering solution to this architectural separation. Extract, Transform, and Load processes (i.e., ETL pipelines) were built to bridge the two worlds, pulling data periodically from operational systems, reshaping it into the dimensional models or flat schemas that analytical systems preferred, and loading it into the data warehouse where analysts and reporting tools could query it. Over time, these pipelines grew in complexity, proliferated across organizations, and became, in many enterprises, the single largest source of data engineering effort, operational overhead, and architectural fragility. A large enterprise data environment might contain hundreds or even thousands of distinct ETL pipelines, each representing a dependency chain between a source system and a downstream consumer, each requiring maintenance as schemas evolved and business logic changed.

The convergence of data environments represents a direct architectural response to the limitations of this model. Converged data platforms (e.g., Oracle 26ai, Snowflake, Google AlloyDB, SingleStore, Databricks with Delta Lake, and MongoDB Atlas) are built on the premise that the technical constraints that necessitated the OLTP/OLAP separation have been sufficiently relaxed by advances in hardware, distributed systems design, and storage architecture that a unified platform can now serve both workload types without the performance degradation and consistency trade-offs that earlier attempts at convergence produced. These platforms do not just co-locate transactional and analytical workloads on shared infrastructure. They fundamentally redesign the storage and query execution layers to serve both efficiently from a single, coherent representation of data.

Oracle's approach to convergence is particularly instructive because it extends beyond the OLTP/OLAP boundary to address the third major dimension of enterprise data: unstructured and semi-structured data content. Oracle 23ai introduced the concept of JSON *Relational Duality*, essentially a design in which the same underlying data can be accessed and manipulated simultaneously as relational tables through SQL or as JSON documents through document API interfaces, without maintaining separate copies or synchronization processes between them. The same row in a relational table is natively accessible as a JSON document, and updates made through either interface are immediately consistent in the other. This duality is not only a convenience feature but represents an architectural statement that the distinction between structured relational data and semi-structured document data (a distinction that drove organizations to build separate document stores alongside their relational databases, creating yet another category of pipeline complexity) is no longer a necessary constraint of the underlying storage system.

How Convergence Transforms the Data Pipeline

The impact of converged data environments on data pipelines is profound and operates across several distinct dimensions. The *first* and most immediate is the elimination of entire categories of pipeline. When transactional and analytical workloads operate on the same live dataset within a single converged platform, the ETL pipelines that previously bridged separate OLTP and OLAP systems become *unnecessary* for internal data movement within that platform. An analytical query against a converged database does not need to wait for an overnight batch load to complete; it reads from the same storage layer that the transactional workload writes to, with isolation mechanisms ensuring that long-running analytical queries do not degrade transactional performance. The pipeline is eliminated not by engineering it away but by removing the architectural gap it was built to bridge.

This elimination is not universal… organizations will continue to need pipelines for integrating data across disparate systems, for ingesting data from external sources, and for moving data between the converged platform and specialized consumers. But the scope and complexity of the pipeline landscape in a well-designed converged architecture is substantially smaller than in a traditional siloed architecture. The pipelines that remain are integration pipelines, not internal movement pipelines, and their design can be focused on the genuinely difficult problem of cross-system integration rather than on the engineering overhead of maintaining consistency between the OLTP and OLAP representations of the same organizational data.

The *second* dimension of impact is pipeline simplification through schema unification. In traditional architectures, one of the most complex and failure-prone aspects of ETL pipeline development was *schema transformation*, essentially the process of mapping the normalized, write-optimized schema of the source OLTP system to the denormalized, read-optimized dimensional schema of the target data warehouse. These transformations were often organizationally fragile, because they embedded business logic in pipeline code that was difficult to version, test, and maintain. When source schemas changed (as they invariably did as applications evolved) pipeline transformations broke, and data quality downstream degraded in ways that were sometimes difficult to detect quickly.

As evident in Table 7-1, converged platforms that support native semi-structured data types alongside relational schemas dramatically reduce this transformation burden. Data that arrives from source systems in JSON format (the native output of most modern APIs and microservices) can be stored, queried, and analyzed within the converged platform without the transformation step that traditional warehousing required. The pipeline's job is reduced to ingestion and routing rather than structural transformation, making it simpler, faster, and less fragile.

Table 7-1: Traditional versus converged architecture

Traditional Siloed Architecture	Converged Architecture
Separate OLTP and OLAP systems requiring ETL bridges	Single platform serving both transactional and analytical workloads
Schema transformation embedded in pipeline logic	Native support for relational, JSON, and unstructured types
Hours-to-days data latency due to batch ETL cycles	Sub-second to real-time data availability for analytics
Hundreds of pipelines creating maintenance sprawl	Integration pipelines only — internal movement eliminated
Separate vector stores required for AI embeddings	Native vector indexing and similarity search within the platform
Multiple governance frameworks across system silos	Unified governance, lineage, and access control in one layer

The *third* dimension of impact is governance consolidation. In traditional multi-system architectures, data governance (i.e., the policies, processes, and technical controls that ensure data quality, security, privacy, and regulatory compliance) had to be implemented separately across each silo. Access controls in the OLTP system had to be replicated in the data warehouse. Data lineage tracking, which records the origin and transformation history of each data element, had to be reconstructed across pipeline boundaries where it was frequently incomplete or inaccurate. Data quality rules applied in the warehouse might not be consistent with those enforced in the source system, creating divergence that was difficult to detect and expensive to resolve.

Converged platforms enable governance to be designed and enforced once, at the platform level, with consistent application across all workload types and all data modalities. Lineage is captured automatically because all data movement occurs within the platform's visibility. Access controls are defined centrally and applied uniformly regardless of whether data is being accessed through a transactional API, an analytical SQL query, or an AI model inference request. This governance consolidation is not solely an operational convenience… it is increasingly a regulatory necessity, as data privacy regulations such as GDPR, the California Consumer Privacy Act, and sector-specific frameworks in financial services and healthcare require organizations to demonstrate comprehensive, auditable control over how personal data is stored, processed, and accessed across their entire data infrastructure.

The Shift from Batch to Streaming: Architectural Implications

The batch processing model that has dominated enterprise data pipelines since the earliest days of business computing reflects a set of assumptions about the temporal relationship between data and decisions that are increasingly misaligned with the realities of modern business operations. In a batch architecture, data is collected over some time period (e.g., an hour, a night, a week) and processed as a discrete unit at the end of that period. The processing job runs, data is transformed and loaded, and analytical consumers have access to an updated snapshot of organizational data that is accurate as of the end of the most recent batch window. Until the next batch cycle completes, the analytical view of the world remains static regardless of how much has changed in the operational systems.

This model was designed for a business environment in which most consequential decisions operated on time horizons that batch latency did not significantly impair. A monthly sales report, a quarterly financial consolidation, an annual strategic planning exercise… none of these were meaningfully degraded by data that was twelve or twenty-four hours old. But the business environment today has changed. The decisions that AI systems are being deployed to support are overwhelmingly time-sensitive. Fraud occurs in milliseconds. Customer churn begins with a single unresolved interaction. Supply chain disruptions cascade within hours. Competitive pricing opportunities open and close in minutes. For AI systems supporting decisions at these timescales, batch-processed data is not merely imperfect… it is structurally unsuitable for the task.

The Architecture of Streaming Pipelines

Streaming pipelines are architecturally distinct from *batch* pipelines (Table 7-2) in ways that go beyond simply processing data more frequently. A batch pipeline is a directed acyclic graph of processing steps that executes periodically: source data is read, transformed through a sequence of operations, and written to a destination, after which the pipeline completes and waits for the next scheduled execution. A streaming pipeline is a continuously running process that consumes data as it is produced, applies transformations within a defined time window or as individual events arrive, and writes results to downstream consumers with latency measured in milliseconds to seconds rather than minutes to hours.

The foundational technology enabling streaming pipelines at enterprise scale is the distributed event streaming platform, of which Apache Kafka has become the dominant standard. Kafka functions as a persistent, distributed message log… a high-throughput, fault-tolerant system that captures every event produced by source systems, retains those events for a configurable period, and makes them available to any number of downstream consumers in the order they were produced. Applications, microservices, databases, and AI inference engines can all publish events to Kafka topics and subscribe to consume them, creating a decoupled, many-to-many data distribution architecture that replaces the point-to-point pipeline connections of traditional ETL with a shared, durable event backbone.

Apache Flink has emerged as one of the leading stream processing engines for applying complex transformations to event streams at scale. Where Kafka handles the transport and persistence of event streams, Flink provides the computational framework for processing them… applying filtering, aggregation, enrichment, joining across multiple streams, and stateful computation that maintains context about each entity across a sequence of events. The combination of Kafka for event transport and Flink for stream processing represents the dominant architecture for enterprise streaming pipelines, though commercial platforms including Confluent, AWS Kinesis, and Google Dataflow offer managed alternatives that reduce the operational complexity of self-managed Kafka and Flink deployments.

Table 7-2: Model characteristics by architectural layer

Architecture Layer	Batch Model	Streaming Model
Data Transport	File-based (SFTP, S3 drops, database exports)	Event streaming (Kafka, Kinesis, Pub/Sub)
Processing Engine	Spark batch jobs, SQL stored procedures, ETL tools	Flink, Spark Structured Streaming, ksqlDB
Latency Profile	Minutes to hours (batch window)	Milliseconds to seconds (event-driven)
State Management	Stateless transformation of complete snapshots	Stateful processing maintaining entity context
Error Handling	Retry full batch on failure	At-least-once or exactly-once delivery semantics
Scaling Model	Scale at job execution time	Continuous autoscaling based on event volume
Data Freshness	Current as of last batch completion	Continuously updated as events arrive

The Rise of Lambda and Kappa Architectures

The practical transition from batch to streaming has not been instantaneous or binary. Most enterprise data environments contain a mixture of use cases, some of which genuinely benefit from streaming and some of which are adequately served by batch processing at lower infrastructure cost. Today, two architectural patterns have emerged to address this reality, each representing a different philosophy about how to manage the coexistence of batch and streaming processing within a single data infrastructure.

The *Lambda architecture*, introduced by Nathan Marz in the early 2010s, addresses the batch-streaming coexistence challenge by maintaining two parallel processing paths for the same data. The batch layer processes the complete historical dataset on a periodic schedule, producing highly accurate but time-delayed views of all data. The speed layer processes the same data stream in real time, producing approximate but current views that cover the period since the last batch cycle completed. Query results are assembled by combining the batch layer's historical view with the speed layer's recent updates, giving consumers a complete and relatively current picture of the data. Lambda architecture solved a real problem at the time of its introduction, but it introduced significant complexity: two separate codebases implementing the same business logic in different frameworks, operational overhead of maintaining two parallel infrastructure stacks, and the non-trivial engineering challenge of correctly merging batch and speed layer outputs at query time.

The *Kappa architecture*, proposed by Jay Kreps as a simplification of Lambda, eliminates the batch layer entirely, treating all data processing as a stream processing problem. Historical reprocessing (the use case that Lambda's batch layer was designed to address) is handled by replaying events from the Kafka event log, which retains events for configurable retention periods ranging from hours to indefinitely. Kappa is architecturally simpler than Lambda, eliminating the dual-codebase problem and the batch/stream merge complexity, but it requires a streaming framework capable of efficiently reprocessing large historical datasets and a Kafka deployment sized to retain sufficient event history for reprocessing needs. As stream processing frameworks have matured and Kafka storage costs have declined, Kappa has become the preferred architecture for greenfield data platform designs where streaming is the primary use case.

The convergence of data platforms is rendering much of the Lambda versus Kappa debate less significant than it once was, because modern converged platforms increasingly handle the integration of streaming and batch workloads within the platform itself. Apache Iceberg, the open table format that is becoming the standard storage layer for the lake house architecture, supports streaming ingestion and batch ingestion against the same table with atomicity, consistency, isolation, and durability (ACID) transaction semantics… meaning a streaming pipeline can be continuously writing events to an Apache Iceberg table while a batch analytical query reads a consistent snapshot of that table, without the two interfering with each other. The distinction between batch and streaming is becoming an implementation detail of the data pipeline rather than a fundamental architectural boundary, as the platforms consuming pipeline output handle both seamlessly.

Micro-Batch as a Pragmatic Middle Ground

Between the extremes of daily batch processing and true event-by-event streaming lies a practically important middle ground: micro-batch processing. In a micro-batch architecture, data is processed in very small, frequent batches (every few seconds, every minute, or every few minutes) rather than in daily or hourly batch cycles or in continuous event streams. Apache Spark Structured Streaming, one of the most widely deployed streaming frameworks in enterprise environments, operates on a micro-batch model by default, processing accumulated events in configurable intervals rather than processing each event individually as it arrives.

Micro-batch architecture offers a pragmatic compromise for organizations transitioning from batch to streaming. It delivers dramatically lower data latency than traditional batch pipelines, reducing data freshness from hours to seconds or minutes, while leveraging the same Spark-based processing infrastructure and SQL-oriented development model that many data engineering teams have already invested in building expertise around. For AI use cases where data currency measured in seconds matters (e.g., personalization, near-real-time fraud scoring, or supply chain alerting) micro-batch can deliver most of the business value of true streaming at a fraction of the architectural complexity and operational cost.

Converged Architecture and AI Platform Accommodation

The difficulty that traditional siloed data architectures create for AI platforms is not primarily a matter of raw computational capability. The algorithms and model architectures that power modern enterprise AI are largely agnostic to the underlying data infrastructure… a gradient boosted tree model for fraud detection, a large language model for contract analysis, or a recommendation engine for personalization can all, in principle, be trained and operated against data drawn from a traditional siloed architecture. The friction is more subtle and more pervasive: it manifests as the cumulative cost in time, engineering effort, data quality, and operational complexity of bridging the gap between where AI models need data and where that data resides in a traditional architecture.

Consider the data preparation burden that traditional architectures impose on AI development. A data scientist building a predictive churn model in a traditional environment must first identify which source systems contain the relevant signals… for example, customer transaction history in the CRM, product usage telemetry in the application database, support ticket data in the service platform, or billing data in the ERP. The data scientist must then work with data engineers to build or extend ETL pipelines to bring this data into the analytical environment. The data arrives in different schemas, at different granularities, with different temporal coverage, and must be joined, cleaned, and transformed into a training dataset. This process (which has nothing to do with model architecture, algorithm selection, or the actual machine learning work) can consume the majority portion of a data science project's time and resources. Studies of enterprise AI programs consistently find that data preparation accounts for sixty to eighty percent of total project effort. The architectural implication is clear: every complexity introduced by siloed data environments translates directly into AI development cost and velocity.

Native Vector Support: The AI-Native Database

The most significant recent development in the convergence of data platforms and AI infrastructure is the native integration of vector storage and vector search capabilities into converged database platforms. To understand why this matters, it is necessary to understand the role that vector embeddings play in modern AI systems. As noted, a vector embedding is a numerical representation of an object (i.e., a document, an image, a customer record, a product description) as a point in a high-dimensional mathematical space, generated by passing the object through a trained neural network. The critical property of embeddings is that objects that are semantically similar are represented as points that are geometrically close together in the embedding space. This property enables a class of AI capability referred to as *semantic* search or *similarity* search: given a query object, find the objects in a dataset that are most semantically similar, regardless of whether they share exact keywords or structured field values.

Semantic search using vector embeddings is the foundational capability underlying RAG, in design the architecture that allows large language models to be grounded in an organization's proprietary data by retrieving relevant documents or records before generating a response. It powers recommendation systems that find products similar to a customer's interest. Semantic search enables anomaly detection systems that identify transactions similar to previously flagged fraudulent patterns; and it supports clinical decision support systems that retrieve patient cases similar to the current patient's presentation. In each of these applications, the AI system needs to store vector embeddings alongside the source data they represent, and to perform fast nearest-neighbor search across potentially billions of vectors to retrieve relevant results at query time.

Traditionally, vector storage and search required a separate, specialized infrastructure component (e.g., a vector database such as Pinecone, Weaviate, or Chroma) sitting alongside the organization's relational database or data warehouse. This separation created exactly the kind of pipeline and synchronization complexity that converged architectures are designed to eliminate: data had to be extracted from the source system, passed through an embedding model to generate vector representations, and loaded into the vector database, creating a separate copy of the data with its own consistency, governance, and freshness challenges.

Oracle 26ai addresses this directly through native vector data type support and integrated vector indexing within the relational database engine. The same table that stores a customer record with structured fields for account balance, transaction history, and demographic attributes can also store a vector embedding of that customer's behavioral profile, generated by an embedding model and stored as a native vector column. SQL queries can combine traditional relational predicates (e.g., filtering by account type, geographic region, or transaction amount) with vector similarity search against the embedded behavioral profile, in a single query executed within the unified platform. This capability, which Oracle markets as *AI Vector Search*, eliminates the vector database as a separate infrastructure component for organizations that adopt the converged platform, bringing vector-based AI capabilities within the same governance, security, and operational framework as all other organizational data.

Snowflake's Cortex AI capabilities, Databricks' Vector Search feature, and Google AlloyDB's "pgvector" integration represent parallel implementations of the same convergence principle across different platform architectures. The *industry direction* is clear: vector storage and search are becoming standard capabilities of the general-purpose converged data platform rather than specialized functionality requiring separate infrastructure.

Real-Time Inference at the Data Layer

Beyond vector search, converged architectures enable a category of AI capability that is simply not achievable within traditional siloed architectures: real-time inference at the data layer, where AI model execution occurs within the database platform itself rather than in a separate model serving infrastructure. This capability (sometimes called in-database machine learning or database-native AI inference) allows AI models to be applied to data at the point where it is stored, rather than requiring data to be extracted from the database and passed to an external model serving endpoint for scoring.

The practical significance of this capability is most evident in high-throughput, low-latency use cases. A fraud detection system that must score every payment transaction as it occurs cannot afford the round-trip latency of extracting transaction data from the operational database, passing it to an external model serving endpoint, waiting for a score, and returning the result to the transaction processing system. Each step in this chain adds latency (e.g., network roundtrips, serialization overhead, queue waiting time) that accumulates into response times measured in hundreds of milliseconds or seconds, which is too slow for a system that must make a fraud decision before a payment authorization response is returned to the merchant. In-database inference eliminates these steps by executing the model directly within the database engine's query execution framework, accessing the transaction data without network overhead and returning a score as part of the transaction processing flow.

Oracle 26ai supports in-database inference through integration with ONNX… the Open Neural Network Exchange format, a standard model representation that allows models trained in frameworks such as PyTorch, TensorFlow, and scikit-learn to be imported into the database and executed as native SQL functions. A model trained by a data science team in a Python notebook can be exported to ONNX format, imported into the Oracle database, and invoked as a SQL function that scores any query result… without requiring the data to leave the database or the organization to operate a separate model serving infrastructure. Snowflake's Cortex ML Functions and BigQuery ML offer analogous capabilities within their respective platform architectures.

Converged Architecture and the Agentic AI Frontier

The accommodation that converged architectures provide for AI platforms is not limited to the model training and inference use cases that characterize current enterprise AI deployments. It extends with particular relevance to the emerging category of agentic AI systems (i.e., autonomous AI agents that plan, reason across multiple steps, use tools, and take actions in pursuit of a goal) which represent, as this book has argued throughout, the most significant near-term frontier of enterprise AI capability.

Agentic AI systems place demands on data infrastructure that are qualitatively different from those of conventional AI models. A conventional predictive model is queried with a fixed input, applies a learned transformation, and returns a fixed output… the data interaction is a simple read operation. An agentic system, by contrast, engages in an extended, multi-step interaction with data: it queries the database to understand the current state of the environment, formulates a plan based on what it finds, takes actions that may modify the environment, observes the results of those actions, and iterates… reading, writing, and reasoning across potentially dozens of data interactions in the course of completing a single task. This pattern of extended, stateful, read-write interaction with live operational data is precisely what converged architectures supporting both transactional and analytical workloads are designed to handle efficiently.

The governance implications of agentic AI further reinforce the value of converged architecture. An agentic system operating with meaningful autonomy across an organization's data must be subject to consistent, enforceable access controls that govern what data it can read, what modifications it can make, and what actions it can take. In a siloed architecture, where the agent might need to interact with a relational database, a document store, a vector database, and a data warehouse executing a single task, implementing consistent governance across these interactions requires coordinating separate security models across separate systems… a complex and fragility-prone proposition. A converged platform that provides a unified governance layer across all data modalities enables the organization to define the agent's access permissions once, enforce them consistently regardless of which data type or query pattern the agent is using, and audit all the agent's data interactions in a single lineage and access log.

Synthesis: Designing Data Pipelines for the Converged Era

The implications of converged data environments, streaming architectures, and AI-native platform capabilities for data pipeline design are substantial and practical. Organizations that approach pipeline architecture with a clear understanding of these shifts will make investment decisions that age well… building infrastructure that accommodates AI capabilities that are emerging now and will mature over the next five years. Organizations that approach pipeline architecture as though the OLTP/OLAP separation and batch processing paradigms are permanent fixtures of the data landscape will build infrastructure that requires expensive rearchitecting precisely when competitive pressure to deliver AI-driven capabilities is highest.

Several design principles follow directly from the analysis in this section. The *first* is that pipeline architecture decisions should be driven by decision latency value, the question of how much business value is generated by data that is more current, rather than by default assumptions about batch processing as the standard approach. Batch processing remains appropriate for many use cases and should not be abandoned for its own sake. But the decision to use batch rather than streaming should be made consciously, with an understanding of what AI capabilities batch latency forecloses, rather than by defaulting to the familiar.

The *second* principle is that pipeline complexity should be minimized by maximizing the scope of the converged platform. Every pipeline represents a point of potential failure, a governance gap, and a maintenance obligation. Pipelines that exist to move data between separate OLTP and OLAP systems within the same organization should be the first target for elimination as converged platform capabilities mature. The pipelines that remain should be those that genuinely require cross-system integration… connecting the converged platform to external data sources, partner data feeds, or specialized systems that fall outside the platform's scope.

The *third* principle is that vector storage and AI inference capabilities should be evaluated as platform selection criteria on equal footing with traditional performance, scalability, and cost metrics. The value of keeping vector embeddings, AI models, and the data they operate on within a single platform. rather than distributing them across separate specialized systems, is sufficiently large, in terms of reduced latency, simplified governance, and reduced operational complexity, to justify meaningful platform selection weight.

Finally, the shift toward converged architectures and streaming pipelines is not merely a technical evolution, it is a change in what data engineers, data architects, and AI platform teams need to know and how they need to collaborate. The data engineer whose expertise is centered on building and maintaining ETL batch pipelines between OLAP and OLTP systems will find that expertise less central as convergence reduces the need for those pipelines. The skills that will be most valuable in the converged, streaming, AI-native data infrastructure landscape are those associated with event-driven system design, stream processing frameworks, data contract definition between producing and consuming systems, vector embedding management, and the governance disciplines that make autonomous AI agents safe to operate at enterprise scale. Understanding this trajectory, not just as a technology selection question but as a workforce development imperative, is among the most important responsibilities of data leadership in the current era of architectural transition.

CHAPTER 8

Scaling AI in the Enterprise: Responsible AI, MLOps, LLMOps, and AIOps

Responsible AI: Ethics, Fairness, and Governance at Enterprise Scale

AS ORGANIZATIONS SHIFT FROM ISOLATED AI prototypes toward enterprise-scale deployment, they encounter a new class of challenges that extend beyond technical refinement or infrastructure performance. These challenges center on fairness, accountability, transparency, and organizational risk. In early experimentation, teams may focus primarily on achieving accuracy or proving feasibility, but as AI systems begin influencing real customers, employees, and market outcomes, the ethical stakes rise sharply. At scale, even small algorithmic defects can create outsized harm, exposing companies to compliance failures, reputational crises, and financial penalties. Understanding and mitigating these risks is therefore essential for any enterprise striving to deploy AI responsibly.

At the core of ethical AI lies the problem of *algorithmic bias*, a phenomenon in which models produce outcomes that systematically disadvantage one group over another. Although bias is often attributed to "black box" algorithms, it almost always originates in the underlying data. Historical bias (also called systemic bias) emerges when the training data reflects existing social inequalities. For instance, if a hiring model learns from the past decisions of a company that historically promoted mostly male engineers, the model may conclude that men are inherently better candidates. This bias does not arise from the model's intentions but from the historical pattern it was asked to replicate.

Another major source of algorithmic bias is representation bias, which occurs when the training data inadequately reflects the full population the model will serve. This issue became widely recognized during early computer-vision deployments, where face recognition models trained predominantly on light-skinned individuals performed poorly with darker-skinned users. The gap was not due to inherent limitations of the algorithms but due to skewed sampling practices that produced uneven exposure to different demographic groups. In enterprises, similar problems arise when customer datasets omit underrepresented regions, age groups, or income brackets.

Bias can also intensify over time through algorithmic feedback loops, a subtle but dangerous dynamic in which a biased prediction alters real behavior and generates new data that reinforces the original bias. Consider a loan approval model that systematically rejects applicants from a particular neighborhood due to incomplete or skewed credit histories. If residents are repeatedly denied credit, the system fails to gather positive repayment data from them, making future models even more restrictive. Without intervention, such feedback loops compound inequities and erode trust, turning bias from an isolated flaw into an institutionalized pattern.

Addressing these risks requires organizations to define what fairness means in each context and to measure it using quantitative criteria. Fairness is multidimensional: some applications demand demographic parity, ensuring similar approval rates across groups, while others prioritize equal opportunity, focusing on consistent error rates rather than consistent outcomes. A bank may tolerate slight differences in approval rates as long as accuracy and risk assessments remain consistent across demographic groups. Meanwhile, a healthcare system using AI to identify high-risk patients may focus on minimizing false negatives across populations to ensure equitable access to treatment. Because no single fairness metric works universally, cross-functional teams must debate and agree on the most appropriate standard before a model goes live.

Even with agreed-upon fairness metrics, organizations still need visibility into how models arrive at their decisions. Explainable AI (XAI) offers the tools and techniques necessary to interpret predictions, providing insight into which features influenced a particular output. This transparency is vital for internal auditors, regulators, customers, and business executives alike. In practice, XAI helps teams trace an adverse model decision (such as an unexpected loan denial) back to specific factors, ensuring accountability. It also aids developers in diagnosing failures, identifying whether a biased input feature, flawed assumption, or data imbalance is responsible for unexpected behavior.

The broader field of AI governance provides the structure for ensuring fairness, accountability, and compliance across the entire AI lifecycle. Governance encompasses the roles, responsibilities, rules, and processes that guide AI development and use. Central to good governance is data oversight, since most failures originate from weak data management. Robust governance ensures that data lineage is clearly documented, that privacy is protected through mechanisms such as anonymization, and that sensitive data is processed in accordance with previously noted laws such as GDPR, CCPA, and HIPAA. Without strong data governance, even well-designed models may violate legal or ethical boundaries.

Beyond data concerns, enterprises must adopt comprehensive risk management frameworks tailored to AI's unique vulnerabilities. Model risk (where a model is simply wrong) poses operational threats when AI outputs directly shape financial, medical, or safety-critical decisions. Meanwhile, security risks such as adversarial attacks introduce new challenges: attackers can manipulate inputs in subtle ways, causing models to make dangerous errors. A classic example is the ability to fool vision systems in autonomous vehicles by placing small stickers on stop signs, causing misclassification. For enterprises deploying AI at scale, defending against such attacks becomes a core responsibility.

Responsible AI is not a separate initiative but rather a philosophy embedded into every phase of the AI project life cycle. Just as organizations consider feasibility and profitability when evaluating new projects, they must also examine ethical and social impacts. Integrating these considerations early prevents costly retrofits or emergency interventions once systems are already in production. It also ensures that risk mitigation is proactive rather than reactive, strengthening both trust and resilience.

Many enterprises now formalize this practice through multidisciplinary AI Review Boards (ARBs), which evaluate new AI use cases through ethical, legal, and technical lenses. These boards typically include data scientists, legal counsel, ethicists, operational leaders, and representatives from affected business units. Their role is not to slow innovation but to ensure that deployment decisions align with organizational values and regulatory obligations. When functioning effectively, ARBs provide clarity and confidence, particularly for high-risk applications such as biometrics, predictive policing, or credit assessment.

Documentation and auditability complete the foundation of responsible AI. Every production model should be accompanied by detailed records describing the business problem it addresses, the fairness and performance metrics it targets, the data sources it relies on, and the assumptions embedded in its algorithms. Documentation also records known limitations so that future teams understand the model's proper use and boundaries. Crucially, monitoring strategies (especially for detecting drift) must be specified in advance to ensure that models remain accurate and fair as real-world conditions evolve.

AI Center of Excellence (CoE) vs. Decentralized Models: How Should a Business Structure Its Teams?

As business organizations move beyond isolated AI pilots and into enterprise-wide deployment, one of the most consequential structural decisions the organizational leadership will face is deceptively simple to articulate but genuinely difficult to execute: where should AI capability live inside the organization? The answer shapes everything downstream... how fast the organization learns, how consistently it governs AI risk, how effectively it attracts and retains technical talent, and ultimately how much business value it extracts from its investments. Two dominant organizational philosophies have emerged in practice. As discussed previously, the *first* is the *AI Center of Excellence*, a centralized model in which AI expertise, tooling, standards, and governance are housed in a dedicated team that serves the broader enterprise. The *second* is the *decentralized or federated model*, in which AI capability is distributed across business units, embedded directly into the teams that own the problems AI is meant to solve. Each model carries distinct advantages, and each carries risks that are predictable enough to plan for if leadership understands the trade-offs before committing to a structure.

The AI Center of Excellence (CoE) model appeals to organizations that are early in their AI maturity journey, or those operating in heavily regulated industries where consistency, auditability, and risk control are non-negotiable. In a CoE structure, a central team typically owns the AI platform infrastructure, model governance standards, vendor relationships, and the organization's AI talent pipeline. Business units submit use cases, the CoE evaluates feasibility and alignment, and approved projects are resourced and executed with shared tooling and oversight. The advantages are significant: the organization avoids duplicating expensive infrastructure, enforces coherent data governance practices, and builds a concentration of expertise that individual business units could never independently sustain. For a regional bank involved in navigating model risk management requirements, or a pharmaceutical company operating under FDA algorithmic validation standards, the CoE model provides the control architecture that compliance demands. The risk, however, is equally significant. CoE models frequently become organizational bottlenecks... well-intentioned centers of expertise that are chronically under-resourced relative to the demand placed on them. Business units grow frustrated by prioritization queues, central teams lose touch with the operational nuances of the problems they are being asked to solve, and the gap between AI potential and business-unit reality widens rather than closes. When the CoE becomes a gatekeeper rather than an enabler, the model begins to undermine the very agility that AI is supposed to deliver.

The decentralized model inverts this logic entirely. Rather than routing AI work through a central function, organizations adopting a federated approach embed data scientists, ML engineers, and AI product managers directly within business units, i.e., in marketing, supply chain, finance, HR, and wherever else AI capability is needed. These embedded teams are close to the domain knowledge, close to the data, and close to the decision-makers who will use AI outputs. Speed of iteration improves dramatically. Business units develop genuine AI fluency because they own the capability rather than consume it as a service. Some of the most advanced AI deployments in consumer retail, logistics, and financial services have been built this way... by teams that understood both the technical architecture and the operational context deeply enough to optimize for both simultaneously. The risks in this model, however, are the mirror image of the CoE's strengths. Without a centralizing function, organizations fragment their data infrastructure, reinvent solutions that already exist elsewhere in the enterprise, and create inconsistent governance practices that expose the organization to regulatory and reputational risk. In the worst cases, decentralized AI teams develop shadow infrastructure (e.g., bespoke models, undocumented pipelines, ungoverned data flows) that creates technical debt and compliance exposure that only becomes visible during an audit or an incident.

The most effective organizations have largely moved past the binary choice between these two models, adopting instead what practitioners increasingly call the "federated CoE" or hub-and-spoke architecture. In this hybrid structure, a lean central team (the hub) owns the things that genuinely benefit from centralization: platform infrastructure, governance frameworks, model risk standards, vendor contracts, and the community of practice that connects AI practitioners across the enterprise. The spokes are embedded teams within each business unit that own execution, domain-specific model development, and the translation of AI capability into business outcomes. The central team's role is not to build everything, but to make it dramatically easier for distributed teams to build well... providing reusable components, approved tooling, governance guardrails, and a talent network that prevents the isolation and attrition that embedded teams in smaller business units often experience. McKinsey, Spotify, and several large financial institutions have published variants of this model, and while the specific configurations differ, the underlying principle is consistent: centralize what scales, decentralize what requires context.

Choosing between these models (or designing the hybrid that fits a specific organization) requires honest assessment along three dimensions. *First, AI maturity*: organizations earlier in the journey typically need the CoE's structure and standards before decentralization can be executed responsibly. *Second, regulatory environment*: industries with strict model governance requirements may never be able to fully decentralize without creating unacceptable compliance risk. *Third, organizational culture*: a highly autonomous, business-unit-driven culture that has historically resisted central functions will struggle to sustain a CoE no matter how well it is designed, and leadership should either invest in the cultural change required or architect a governance-light federated model that works with the grain of the organization rather than against it. There is no universally correct answer, but there is almost always a wrong answer for a specific organization… and it is usually the one chosen without these trade-offs being explicitly surfaced. The structure of AI teams is not a *technical* decision. Rather, it is a *strategic* one, and it deserves the same rigor that any major organizational design decision receives.

Role Ambiguity and Displacement: Managing Workforce Anxiety and the Psychological Barriers to AI Adoption

No technology in modern business history has arrived with quite the same psychological weight as artificial intelligence. Previous waves of automation (e.g., robotic process automation, enterprise resource planning systems, even the introduction of personal computing) displaced specific categories of task and role, but they did so within boundaries that most workers could intuitively understand. A factory floor worker could see the robotic arm that replaced a welding function. A travel agent could identify the moment Expedia made their role economically unviable. AI is *different* in a way that is deeply unsettling to the workforce encountering it, because its boundaries are not visible, its ceiling is not yet known, and its encroachment into cognitive work (the domain that educated professionals have long considered their *protected* territory) feels qualitatively different from anything that preceded it. The result is a form of workforce anxiety that is neither irrational nor easily resolved, and organizations that dismiss it as resistance to change, rather than treating it as a legitimate psychological and professional response to genuine uncertainty, will pay for that dismissal in adoption failures, talent attrition, and cultural fracture that no technology investment can overcome.

The concept of role ambiguity in the context of AI adoption refers to the condition in which an employee can no longer clearly define the boundaries of their own job function, the criteria by which their performance will be evaluated, or their long-term relevance within the organization. This ambiguity operates at multiple levels simultaneously. At the task level, workers find that AI tools have absorbed specific activities that previously justified significant portions of their time (e.g., drafting, summarizing, analyzing, coding, and classifying) without any corresponding clarification from leadership about what is now expected of them in the hours freed. At the identity level, professionals whose expertise and organizational status were built on mastery of exactly those activities face a more profound disruption: if the knowledge I spent a decade acquiring can now be approximated by a language model in seconds, what is the nature and value of my expertise? At the strategic level, employees watch their organizations announce AI transformation initiatives without accompanying clarity about which roles will evolve, which will be eliminated, and what will be the timeline and criteria for those decisions. Research in organizational psychology consistently demonstrates that ambiguity of this kind (particularly when it involves perceived threat to status, identity, and economic security simultaneously) produces *anxiety* responses that are more debilitating and longer lasting than responses to clearly communicated bad news. Employees can adapt to a definitive answer they dislike. They struggle profoundly to function effectively inside sustained uncertainty with no resolution in sight.

The psychological barriers that emerge from this environment manifest in patterns that AI project leaders frequently misdiagnose. Passive non-adoption (i.e., employees who complete required AI training, acknowledge the tools in meetings, and then quietly continue working in their prior methods) is often attributed to technical discomfort or insufficient training. In most cases it is neither, but rather a rational hedging strategy by workers who have not been given sufficient reason to trust that their investment in AI capability will be recognized and rewarded rather than used as justification for their eventual replacement. Active resistance, when it surfaces, is similarly misread as technophobia or generational stubbornness. More often it reflects a legitimate grievance: that the organization is asking workers to participate enthusiastically in a transformation whose benefits will accrue primarily to shareholders and executives, while the risks, namely displacement, deskilling, or loss of bargaining power, will be borne disproportionately by the workforce. Perhaps the most organizationally damaging pattern is the quiet departure of the organization's most capable and mobile employees (precisely those with the skills and market options to leave) who conclude that the organization's AI trajectory makes their long-term prospects there less attractive than alternatives elsewhere. This situation represents a particularly cruel irony: AI adoption efforts, if poorly managed, can trigger the talent exodus that makes sophisticated AI adoption impossible.

Addressing these dynamics requires a change management philosophy that is substantively different from the communication campaigns most organizations default to. Announcing that "AI will *augment*, not replace" human workers is not a strategy… it is a slogan, and a workforce that has watched peers displaced in other industries or other organizations will receive it as such. What actually moves the needle is behavioral specificity from leadership: concrete articulation of which roles are being redesigned and how, transparent criteria for how AI performance will be evaluated alongside human performance rather than instead of it, visible investment in reskilling that is meaningful enough to represent a genuine career development commitment rather than a compliance exercise, and organizational structures that give employees agency in how AI tools are adopted within their domains. Psychological safety (i.e., the condition in which employees can raise concerns about AI adoption, report failures, and question design decisions without fear of being labeled as obstructionist) is not a soft cultural amenity in this context. It is a hard *operational* requirement. Organizations that create it will surface the failure signals that allow them to course-correct before projects fail. Those that suppress it will discover those failure signals only in retrospect, when the cost of correction is far higher. Managers at every level need to be equipped not just with AI literacy, but with the conversational and coaching skills to hold honest, ongoing dialogue with their teams about what AI transformation means in a practical sense for the people doing the work.

The concept of the "*organizational ceiling*", meaning the point at which human and cultural friction becomes the binding constraint on AI value realization, regardless of the quality of the underlying technology is nowhere more visible than in the dynamics of role ambiguity and displacement anxiety. Organizations consistently underestimate the impact of this ceiling because it is invisible in the project plan, unmeasured in the ROI model, and uncomfortable to discuss in leadership forums where optimism about transformation is the expected register. However, the research evidence, and the accumulating record of enterprise AI initiatives that stalled not for technical reasons but for human ones, is unambiguous: the psychological dimension of AI adoption is not a secondary concern to be addressed after the technology is deployed. It is a primary success condition that must be designed for from the beginning, resourced with the same seriousness as the model infrastructure, and measured with the same rigor as the technical KPIs. The business enterprises that will lead in the AI era are not necessarily those with the most sophisticated models. They are those that figured out how to bring their people (genuinely, not rhetorically) along for the transformation.

AI Change Management: Overcoming the "Human Ceiling" and Psychological Resistance

Every significant technology transformation in modern business history has ultimately been a human problem wearing a technical costume. The organizations that successfully navigated the transition to enterprise resource planning systems in the 1990s, the shift to cloud infrastructure in the 2000s, and the adoption of agile operating models in the 2010s *did not succeed* primarily because they chose better software or hired more capable engineers. They succeeded because they understood that technology adoption is fundamentally a *change management* challenge, and they invested accordingly in the human architecture required to make new ways of working stick. Artificial intelligence represents this dynamic in its most acute form yet. The gap between what AI systems are technically capable of delivering and what organizations substantively realize from their AI investments is (in the majority of documented cases) not explained by model performance, data quality, or infrastructure limitations. It is explained by what was referred to prior as the "human ceiling"… the invisible but very real boundary above which AI value creation cannot rise when the human systems surrounding the technology are not designed, led, and managed with the same rigor as the technology itself. Understanding this ceiling, diagnosing its specific manifestations inside a given organization, and building the enterprise change management architecture to raise or remove it entirely is among the most important and most neglected capabilities in enterprise AI leadership today.

To drill down, the human ceiling is not a single phenomenon but a composite of overlapping psychological, cultural, and structural forces that interact in ways that can be difficult to disentangle in practice. At the individual level, it begins with what psychologists call loss aversion… the well-documented human tendency to weight potential losses more heavily than equivalent potential gains. When an employee encounters an AI tool that promises to make them more productive, their cognitive calculus is not simply a rational assessment of the upside. It is a simultaneous, often unconscious assessment of what adoption might cost them: the tacit knowledge that currently makes them valuable, the workflow autonomy they have spent years earning, the collegial relationships built around shared practices that AI adoption might restructure, and the fundamental question of whether demonstrated AI proficiency will be recognized as a professional asset or used as a benchmark that raises expectations without raising compensation. This loss aversion is compounded by what organizational behaviorists call identity threat… the disruption that occurs when a technology challenges not just how a person works but who they understand themselves to be professionally. For example, a senior financial analyst whose identity is built around the sophisticated interpretive judgment they bring to complex data faces a categorically different psychological challenge than a junior analyst being asked to use a new reporting tool. For the senior professional, AI adoption can feel less like an efficiency upgrade and more like an implicit argument that their years of accumulated expertise are worth less than they believed. Organizations that do not design their organizational change management approach to address this identity dimension will find that no amount of training, communication, or incentive structure fully resolves the resistance they encounter.

At the team and middle-management level, the human ceiling manifests through a distinct and particularly damaging set of dynamics. Middle managers occupy a structurally precarious position in AI transformation: they are simultaneously responsible for driving adoption among their teams and acutely aware that AI-driven process changes frequently reduce the coordination and oversight functions that justify their own roles. This situation creates a conflict of interest that is rarely acknowledged openly but is almost universally present, and it produces a management behavior that researchers have called *performative compliance*... managers who visibly endorse AI initiatives in leadership forums while quietly failing to prioritize adoption in their day-to-day team management, deprioritizing AI tool usage in performance conversations, and allowing the path of least resistance to remain the path of prior practice. This dynamic is *not* malicious sabotage, but rather a rational, self-protective response to structural ambiguity, and it will persist until the organization resolves that ambiguity by making clear what AI transformation means for the management layer specifically... including honest acknowledgment of where management roles will change, what new capabilities will be valued, and what the development pathway looks like for managers who invest genuinely in building AI fluency. Ignoring this dynamic, as many AI transformation programs do by focusing change management energy exclusively on frontline workers, is one of the most reliable predictors of program stall at the scaling stage.

Effective enterprise AI change management requires a framework that operates simultaneously across three distinct but interdependent levels. The *first is the cognitive level*... ensuring that employees have not just technical training in AI tools but a sufficiently accurate mental model of what AI really involves, what it can and cannot do, and how its role in the organization will evolve over time. Anxiety thrives in information vacuums, and the popular discourse around AI (saturated with both utopian and apocalyptic narratives) actively fills those vacuums with frameworks that are rarely helpful and often paralyzing. Organization leaders have a responsibility and a strategic interest in replacing that ambient noise with grounded, specific, honest communication about the AI tools being deployed, the use cases being prioritized, the limitations that have been identified, and the process by which employee input will be incorporated into ongoing development. The *second level is the emotional level*... acknowledging explicitly, in leadership communication and in manager-employee dialogue, that the feelings employees are experiencing about AI transformation are legitimate responses to genuine uncertainty rather than symptoms of a training deficit to be corrected. This acknowledgment costs nothing and unlocks an enormous amount of psychological energy that is currently being spent on anxiety management rather than productive engagement. The *third level is the structural level*... redesigning the incentive systems, performance frameworks, role definitions, and career pathways that currently make AI adoption feel risky at the individual level. If employees are evaluated on the outputs they personally produce rather than the outcomes their team achieves, AI tools that make their individual contribution less visible will rationally be resisted regardless of how compelling the organizational business case appears.

The organizations that have most successfully navigated these dynamics share a set of practices that are worth examining as a practical template. They appoint change leaders (not just project managers) who are specifically accountable for the human adoption dimensions of AI transformation, with standing in leadership forums equivalent to that of the technical program leads. They build feedback mechanisms that give frontline employees genuine influence over how AI tools are configured and deployed in their specific work contexts, creating the sense of agency that is the single most reliable predictor of adoption quality. They stage their transformations in ways that allow employees to accumulate visible wins with AI tools before the stakes of adoption are raised, building the confidence and competence that makes deeper integration feel like progression rather than threat. Moreover, they treat organizational change management not as a communications campaign that runs alongside the technical program but as a design discipline that begins at the earliest stages of use case selection and continues through post-deployment optimization. The human ceiling is not a fixed architectural feature of any organization. It represents a leadership and management variable… one that rises or falls in direct proportion to the quality of attention, investment, and genuine human respect that the organization brings to the people who must ultimately make AI work in practice. The technology will continue to improve regardless of what any single organization does. The competitive differentiator, increasingly, is not who has access to the best models but rather who has built the organizational capability to effectively use them.

Structuring for Success: Centralized vs. Federated AI Organizational Models

The question of how to structurally organize AI capability inside an enterprise is one that every organization pursuing serious AI transformation will eventually be forced to answer, and the consequences of answering it poorly are both significant and slow to reveal themselves. Unlike a failed pilot project, which surfaces its inadequacy quickly and at contained cost, a poorly chosen organizational model for AI embeds its dysfunction into the operating architecture of the enterprise… shaping hiring decisions, governance practices, technology investments, and cultural norms in ways that compound over time and become progressively harder to unwind. The decision is made more difficult by the fact that there is no universally correct answer. The optimal structure for a global financial services firm with tens of thousands of employees, a mature data infrastructure, and a heavily regulated operating environment looks fundamentally different from the optimal structure for a mid-market technology company with a nimble culture, a single core product, and an existential competitive incentive to move fast.

What is universal, however, is the need to make this decision deliberately, with a clear understanding of the trade-offs each model carries, the organizational conditions under which each performs well, and the specific failure modes that each is prone to when those conditions are not present. Structural decisions made by default (because they mirror what a peer organization announced, because they replicate the org chart logic of a prior technology function, or because they avoid the political difficulty of a more intentional choice) are among the most expensive mistakes an AI transformation program can make.

The centralized model, in its purest form, consolidates AI strategy, talent, tooling, governance, and execution authority within a single organizational unit… previously noted as an *AI Center of Excellence*; also an *AI Division* or a *Chief AI Officer organization*. The intellectual appeal of this model is considerable and grounded in legitimate organizational logic. Centralization creates economies of scale in infrastructure investment, preventing the expensive redundancy that occurs when individual business units independently procure, build, and maintain separate AI platforms. It establishes a single point of accountability for AI governance and risk management, which is particularly valuable in regulated industries where model risk documentation, audit trails, and explainability requirements demand consistent standards across the enterprise. Moreover, it enables the concentration of scarce and expensive AI talent in a team with sufficient critical mass to attract senior practitioners, support professional development, and build the institutional knowledge that sustains capability over time rather than allowing it to walk out the door with individual contributors. Centralization also creates the organizational authority to enforce data standards, prevent shadow AI development, and ensure that the enterprise's AI investments are coordinated toward strategic priorities rather than fragmented across disconnected tactical experiments. For organizations in financial services, healthcare, defense contracting, and other sectors where a single AI failure carries regulatory, legal, or reputational consequences that dwarf the cost of the AI program itself, these advantages are not merely convenient… they are often prerequisites for operating responsibly in the space overall.

The limitations of the centralized model, however, are equally structural and equally predictable. Centralized AI organizations face an inherent scaling problem: the demand for AI capability across a large enterprise almost always grows faster than a central team can responsibly serve it, creating a prioritization backlog that frustrates business units, slows value realization, and generates the organizational resentment that eventually undermines the central team's authority and mandate. More fundamentally, centralized teams are structurally distant from the domain knowledge, operational context, and day-to-day decision-making that determines whether an AI application persistently solves a real business problem or simply demonstrates technical sophistication in isolation from business relevance. A central data science team that builds a demand forecasting model for the supply chain function without embedded understanding of how procurement managers actually interpret and act on forecast data will frequently produce technically impressive work that generates minimal operational value not because the model is wrong but because the *organizational distance* between the builders and the users prevented the kind of iterative, contextually grounded collaboration that separates genuinely useful AI from expensive proof-of-concept theater. When centralized AI teams repeatedly encounter this dynamic, the consequences are predictable: business units begin routing around the central function, building informal capability through vendor relationships and shadow investments, and the central team's relevance erodes even as its formal authority remains nominally intact.

The federated model addresses these limitations directly by distributing AI capability into the business units where domain knowledge, operational context, and decision-making authority already reside. In a fully federated structure, each major business function (i.e., marketing, finance, supply chain, human resources, product development, etc.) builds and owns its own AI capability, staffed by practitioners who are embedded in the team, accountable to business unit leadership, and evaluated primarily on the business outcomes their AI work produces rather than on technical metrics that may have limited correlation with actual value creation. The advantages of this proximity are not trivial. Embedded AI teams develop the kind of deep domain fluency that allows them to identify high-value use cases that would never surface through a centralized intake process, to design solutions that fit the actual workflow rather than the idealized workflow, and to iterate in response to user feedback at the speed that operational relevance requires. They also build the business unit's AI literacy organically (through daily collaboration rather than periodic training events) creating a broader organizational capability that makes AI adoption self-sustaining rather than dependent on the central function's bandwidth. Companies that have built significant competitive advantage through AI (e.g., Amazon in logistics and personalization, Netflix in content recommendation, JPMorgan Chase in fraud detection and risk modeling) have in large part done so by embedding AI capability close enough to the operational reality that the models could be continuously refined in response to real-world feedback rather than periodic centralized review cycles.

The federated model's failure modes, however, are the precise inverse of its strengths. Distributed AI development, without strong coordinating mechanisms, reliably produces infrastructure fragmentation… each business unit building its own data pipelines, model training environments, and deployment infrastructure, creating a proliferation of technical debt that becomes increasingly expensive to maintain and increasingly difficult to govern. Governance inconsistency is perhaps the most serious risk: when AI model development standards, bias testing protocols, explainability documentation requirements, and deployment approval processes vary across business units, the organization's aggregate AI risk exposure may be far higher than any individual unit's risk management practices would suggest. Data sovereignty violations, model failures with discriminatory impact, and regulatory compliance gaps frequently originate not in centralized AI programs where governance attention is concentrated, but in federated deployments where the combination of business unit autonomy and technical complexity creates blind spots that neither the business unit nor the central governance function has sufficient visibility to detect. Talent fragmentation is an additional structural vulnerability: embedded AI practitioners in smaller business units may lack the peer community, technical challenge, and career development infrastructure to remain engaged over time, creating attrition patterns that leave business units periodically without the capability on which they have come to depend.

The resolution that most mature AI organizations have converged on involves a hybrid architecture that attempts to capture the governance and scale advantages of centralization while preserving the domain proximity and execution speed of federation. This model (sometimes described as the hub-and-spoke model, the federated CoE, or the platform-plus-embedded-teams architecture) operates on a principled division of responsibility between a lean central function and distributed business unit teams. The central function, or hub, owns the enterprise AI platform infrastructure, the governance and risk management framework, the vendor and partner relationships, the AI talent community of practice, and the cross-enterprise data standards that make interoperability and consistent governance possible. Critically, the hub's role is explicitly defined as an enablement function rather than an execution function… its success is measured not by how many models it builds directly, but by how effectively it raises the capability, speed, and governance quality of the distributed teams it supports. The embedded teams, or spokes, own use case identification, domain-specific model development, business stakeholder relationships, and the translation of AI capability into measurable business outcomes. They operate with genuine autonomy within the guardrails the hub establishes, and they contribute to the community of practice the hub maintains… sharing learnings, reusable components, and failure analyses that make the collective AI capability of the organization greater than the sum of its parts.

Choosing the right point on the centralization-federation spectrum (and designing the specific governance and coordination mechanisms that make the chosen model work in practice) requires leadership to conduct an honest organizational assessment across several dimensions before committing to a structure. *AI maturity is the first*: organizations without established data infrastructure, documented governance standards, or a critical mass of internal AI talent typically need the discipline and coordination that a more centralized starting point provides, even if the long-term target state is a federated architecture. *Regulatory environment is the second*: the higher the regulatory stakes of AI model failures, the stronger the argument for governance mechanisms that centralization makes easier to enforce. *Organizational culture is the third* and frequently the most decisive: an enterprise with a long history of strong business unit autonomy, federated P&L accountability, and cultural resistance to central functions will struggle to sustain a heavily centralized AI model regardless of its theoretical advantages, and leadership should either make the cultural investment required to support centralization or design a federated architecture with governance mechanisms strong enough to manage the associated risks. The *fourth dimension is competitive urgency*: organizations facing existential AI-driven disruption in their core markets may need to accept the governance risks of rapid federated deployment to move at the speed their competitive situation demands, treating governance improvement as an iterative priority rather than a prerequisite for action. In general, there is no correct answer that applies across all dimensions simultaneously. There is, however, always an answer that fits a specific organization's specific combination of maturity, regulatory context, culture, and competitive reality… and finding it is an act of strategic leadership, not technical administration.

MLOps, LLMOps, and AIOps: Operationalizing AI at Enterprise Scale

Machine Learning Operations (MLOps), again refers to the set of practices, processes, and tools designed to manage the end-to-end lifecycle of machine learning models in production. While traditional software development follows well-established DevOps practices, deploying machine learning introduces unique challenges because models are not static code… they are data-driven artifacts that evolve as they are retrained with new data. MLOps aims to bridge the gap between *data science and operational engineering*, ensuring that models can be deployed, monitored, and updated reliably and at scale.

At its core, MLOps involves model versioning, automated testing, continuous integration, continuous deployment, and monitoring. Versioning tracks both the model code and the datasets used to train it, enabling reproducibility and accountability. Automated testing ensures that models meet predefined accuracy, fairness, and safety metrics before deployment, preventing regressions or biased outputs. Continuous integration and deployment pipelines streamline the process of moving models from development to production, integrating with existing IT infrastructure while reducing manual intervention. Once deployed, MLOps emphasizes *real-time monitoring and observability*, tracking model performance, detecting drift in data or predictions, and triggering retraining or rollback as necessary.

Advanced MLOps frameworks also support feature stores, automated retraining, and deployment orchestration across cloud or on-premises environments, integrating seamlessly with Continuous Integration / Continuous Deployment (CI/CD) pipelines. From a managerial perspective, MLOps is critical because it transforms machine learning from a research experiment into a repeatable, scalable, and auditable business capability, reducing operational risks, ensuring regulatory compliance, and maximizing the ROI of AI initiatives. In essence, MLOps is the discipline that turns the promise of AI into reliable, production-grade applications that can continuously learn and adapt in the real world. CI is the practice of frequently integrating code changes into a shared repository. Each update is automatically tested and validated to catch errors early. In MLOps, this action can include validating data pipelines, testing feature engineering code, checking model training scripts, and running unit or integration tests on models. CD automates the process of deploying changes into staging or production environments once they pass CI checks. In an MLOps pipeline, this may involve automatically deploying updated models, refreshing features, rolling out new inference services, or retraining and redeploying models based on performance thresholds.

Large Language Model Operations (LLMOps) referring to an emerging discipline that extends the principles of MLOps specifically to large language models (LLMs), which are generative AI systems capable of understanding, reasoning, and producing human-like text. While traditional MLOps focuses on training, deploying, and maintaining predictive models, LLMOps addresses the unique challenges posed by models that are orders of magnitude larger, require specialized infrastructure, and often interact directly with users in real time.

At its core, LLMOps encompasses model lifecycle management, prompt engineering, fine-tuning, safety alignment, monitoring, and cost optimization. Large language models are typically pretrained on massive datasets and then fine-tuned on domain-specific data to deliver business-relevant outputs. LLMOps ensures that these models are properly versioned, with full tracking of the pretraining datasets, fine-tuning data, and hyperparameters, enabling reproducibility and governance. Unlike traditional ML models, LLMs are often deployed in *multi-modal or conversational environments*, which requires continuous testing for accuracy, relevance, safety, and ethical compliance, particularly to prevent hallucinations, biases, or inappropriate outputs.

A major component of LLMOps involves *prompt engineering, management and optimization*. Because LLM behavior is heavily influenced by prompts, LLMOps frameworks treat prompt templates, chains, and workflows as first-class artifacts. This includes experimentation with prompt phrasing, *RAG* pipelines, and reinforcement learning from human feedback (RLHF) to align outputs with organizational objectives. Monitoring and observability are also crucial, tracking metrics such as latency, token usage, hallucination rate, user satisfaction, and alignment with compliance standards.

Infrastructure and cost management are equally central. LLMs demand *highly parallelized* compute (GPUs or TPUs), significant memory, and fast storage for embeddings, and LLMOps practices include efficient deployment strategies like model *quantization, sharding, and edge inference* to reduce latency and operational costs. Security and data privacy are also baked into LLMOps pipelines, ensuring sensitive information used for fine-tuning or interaction is protected.

From a managerial perspective, LLMOps turns powerful but complex language models into reliable, auditable, and scalable business tools, capable of powering chatbots, virtual assistants, content generation, knowledge retrieval, and decision support. It is the discipline that ensures generative AI systems can be safely integrated into workflows while continuously learning, adapting, and delivering value without overwhelming operational teams or resources (see Table 8-1 for a comparison of MLOps and LLMOps.)

Table 8-1: MLOps and LLMOps share a common ancestry in machine learning operations discipline, but the scale, non-determinism, and generative nature of large language models require distinct tooling, monitoring, and governance practices at nearly every stage of the lifecycle

Aspect	MLOps (Machine Learning Operations)	LLMOps (Large Language Model Operations)	Key Difference / Implication
Model Type	Traditional predictive models (regression, classification, small-scale neural networks)	Large language models, often transformer-based, generative, and multi-modal	LLMs are orders of magnitude larger and generative; MLOps handles smaller, more deterministic models.
Training Data	Structured or semi-structured	Massive unstructured	LLMOps requires data

Aspect	MLOps (Machine Learning Operations)	LLMOps (Large Language Model Operations)	Key Difference / Implication
	datasets; smaller volumes	datasets (text, code, multi-modal), often spanning billions of tokens	versioning, filtering, alignment, and careful handling of sensitive information.
Compute Requirements	Moderate CPU/GPU usage; can run on single machines or small clusters	Extreme compute: multi-GPU/TPU clusters, high memory, fast I/O, sometimes distributed across data centers	LLMOps demands specialized infrastructure planning, cost monitoring, and efficiency optimization.
Deployment	Batch predictions or real-time APIs; endpoints serve deterministic outputs	Often deployed as interactive, generative systems; supports conversational, retrieval-augmented, or real-time multi-modal applications	LLMOps focuses on latency, token usage, streaming responses, and orchestration of RAG pipelines.
Versioning & Lifecycle	Standard model versioning; retrain periodically with new data	Complex versioning of base models, fine-tuned models, prompts, embeddings,	LLMOps treats prompt templates and knowledge bases as first-class artifacts in lifecycle management.

Aspect	MLOps (Machine Learning Operations)	LLMOps (Large Language Model Operations)	Key Difference / Implication
		and RLHF adjustments	
Monitoring & Observability	Metrics such as accuracy, precision, recall, F1 score, and inference latency	Metrics include hallucination rate, relevance, user satisfaction, alignment, token usage, and bias detection	LLMOps requires qualitative and quantitative monitoring because outputs are generative and less deterministic.
Safety & Compliance	Data privacy, bias mitigation, and model fairness	Enhanced safety: hallucination prevention, harmful content filtering, RLHF alignment, privacy-preserving fine-tuning	LLMOps adds continuous ethical and safety supervision.
Optimization Techniques	Hyperparameter tuning, feature engineering, model compression	Prompt engineering, quantization, sharding, retrieval augmentation, fine-tuning on domain-specific data	LLMOps emphasizes prompt workflows and resource-efficient inference at scale.
Team Skills Required	Data engineers, ML engineers, data scientists	Extended team: MLOps engineers,	LLMOps requires broader cross-functional

Aspect	MLOps (Machine Learning Operations)	LLMOps (Large Language Model Operations)	Key Difference / Implication
		prompt engineers, ML safety specialists, AI product managers	collaboration due to complexity and generative nature.

AI Operations (AIOps) refers to the practice of applying artificial intelligence, machine learning, and advanced analytics to optimize, automate, and manage IT operations and complex digital systems. In modern enterprises, IT environments have become extraordinarily complex, spanning cloud platforms, hybrid infrastructures, microservices architectures, containers, and thousands of interconnected applications. Traditional monitoring and management approaches, which rely heavily on manual processes and static rules, are no longer sufficient to maintain reliability, performance, and security at scale. AIOps addresses this gap by ingesting and analyzing massive volumes of operational data (including logs, metrics, events, traces, and alerts) in real time to detect patterns, predict issues, and automate responses.

At its core, AIOps combines three major components:

1. *Data Ingestion and Normalization:* AIOps systems pull data from diverse sources (e.g., application performance monitoring tools, cloud infrastructure logs, network devices, and security alerts) and normalize it into a consistent, structured format suitable for analysis. This component ensures that signals from heterogeneous sources can be interpreted holistically.
2. *Intelligent Analytics and Event Correlation:* Using machine learning algorithms, AIOps identifies relationships and anomalies in operational data. For example, it can correlate seemingly unrelated alerts across services to detect a single underlying incident or predict a server failure before it occurs. This predictive capability transforms IT operations from reactive problem-solving to proactive system management.

3. *Automation and Remediation:* Beyond detection, AIOps enables automated responses to common issues. It can trigger scripts, restart services, reallocate resources, or notify the relevant teams with actionable insights. Over time, the system learns from past incidents, continuously refining its decision-making to reduce downtime and improve efficiency.

From a managerial perspective, AIOps is a *force multiplier*: it reduces operational costs, accelerates problem resolution, and enhances service reliability, all while allowing human IT teams to focus on strategic initiatives rather than firefighting routine alerts. Importantly, AIOps is not limited to IT infrastructure… it can also be applied to AI model operations, monitoring model drift, performance degradation, and resource utilization to ensure AI systems remain accurate, fair, and cost-efficient in production.

Simply put, AIOps transforms IT operations from a reactive, manual function into a data-driven, predictive, and self-optimizing system, leveraging AI to ensure that both traditional infrastructure and AI-powered applications operate seamlessly and efficiently.

In the end, scaling AI responsibly allows organizations to unlock transformative value while preserving trust, reducing risk, and meeting regulatory expectations. The most successful enterprises recognize that ethical considerations are not a barrier to AI innovation but a catalyst for sustainable growth. By embedding fairness, transparency, and governance into their AI operations, they ensure that advanced technologies strengthen, rather than compromise, the integrity of their decisions and the well-being of the communities they serve.

CHAPTER 9

AI-Powered Functional Digital Transformation

From Isolated Pilots to Enterprise-Wide Transformation

ARTIFICIAL INTELLIGENCE IS RESHAPING the way organizations operate across every functional domain, namely marketing, sales, operations, finance, human resources, research, legal, and information technology. What began in most enterprises as a set of isolated proof-of-concept pilots has matured, in the most advanced organizations, into systematic functional transformation: the integration of AI into the core processes, decision workflows, and operational architectures that determine organizational performance. This chapter examines how that transformation unfolds in practice, function by function, and how the integration of AI across functions creates compound value that exceeds the sum of individual deployments.

The distinction between AI experimentation and AI-driven functional transformation is consequential. Experimentation produces learning; transformation produces durable competitive advantage. The difference lies not in the sophistication of the underlying technology but in the degree to which AI is embedded into operational workflows, connected to proprietary data assets, governed by accountability frameworks, and continuously improved through feedback loops that align model performance with business outcomes. Business organizations that have achieved functional transformation (as distinct from those still cycling through pilots) demonstrate measurably superior performance across a range of operational and financial metrics.

This chapter is structured around the primary functional domains in which AI is driving transformation, followed by an examination of the cross-functional integration mechanisms that amplify functional-level impact. For each domain, the analysis covers the AI capabilities being deployed, the organizational and data requirements for effective deployment, the measurement frameworks used to assess impact, and the implementation challenges that organizations most commonly encounter. The chapter concludes with a treatment of the governance, culture, and organizational design considerations that determine whether functional AI transformation delivers sustained value.

Marketing, Sales, and Customer Experience: The Shift from Mass to Intelligent Personalization

Marketing and customer experience have undergone a fundamental transformation as AI has moved the industry from mass communication to individualized engagement at scale. The classical marketing challenge (i.e., matching the right message to the right customer at the right moment) was historically constrained by the cost of personalization and the limitations of available customer data. AI resolves both constraints simultaneously: machine learning algorithms can process behavioral signals at the individual level across millions of customers in real time, and the cost of delivering personalized communications at scale has fallen to near zero.

The practical manifestation of this capability extends well beyond email subject line optimization or basic recommendation engines. Advanced organizations deploy multi-dimensional personalization systems that simultaneously adapt content, channel, timing, offer structure, and creative assets based on individual customer behavioral profiles, purchase history, real-time context, and predicted lifetime value. The outputs of these systems are continuously validated against observed behavioral responses, enabling rapid iteration and improvement. Netflix and Amazon, for example, have demonstrated the commercial power of this approach at scale (their *recommendation systems* are estimated to drive a substantial fraction of total consumption and revenue, respectively) but the underlying personalization architectures are increasingly accessible to organizations of all sizes through platform APIs and enterprise marketing technology.

Beyond personalization at the point of interaction, AI enables organizations to manage customer relationships through the full lifecycle with a predictive intelligence that was previously available only in aggregate. *Churn prediction models* identify customers at elevated risk of attrition before behavioral signals are visible to human account managers, enabling proactive intervention "at the moment" when retention investment has the highest expected return. *Customer lifetime value models* score individual customers on their predicted long-run revenue contribution, enabling differentiated service levels, acquisition investment, and retention prioritization that optimize the economics of the customer portfolio rather than treating all customers equivalently.

Next-best-action systems represent a particularly powerful application of predictive customer analytics. These systems synthesize information from across the customer data platform (e.g., behavioral history, demographic profile, product holdings, service interactions, and predictive scores) to recommend the specific action most likely to maximize customer value at each interaction point, across every channel simultaneously. The sophistication of next-best-action systems has advanced rapidly with the availability of large language models that can generate personalized communication content rather than merely selecting from pre-authored templates, enabling mass customization of customer interactions at a depth that manual processes cannot approach.

AI in Sales

Sales functions benefit from AI across the full sales cycle, from lead generation and qualification through pipeline management and close. AI-driven lead scoring models assess the probability that a given prospect will convert to a customer, allowing sales teams to concentrate their finite time on the opportunities with the highest expected return. These models have evolved substantially beyond simple demographic scoring: modern lead intelligence systems incorporate behavioral signals (e.g., website engagement patterns, content consumption, email response timing) as well as firmographic data, technographic data indicating current technology stack, and predictive signals derived from the behavior of comparable buyers.

Pipeline intelligence systems apply similar predictive logic to existing pipeline management, identifying deals at risk of delay or loss before the signals are apparent to the responsible account executive. By providing early warning of pipeline risk, these systems enable timely coaching interventions, resource reallocation, and escalation decisions that improve forecast accuracy and close rates simultaneously. Natural language processing tools (including conversational AI systems that handle initial prospect engagement, and call analytics platforms that analyze sales conversations for coaching insights) further extend AI's contribution to sales productivity.

AI-powered sentiment analysis and voice-of-customer systems provide organizations with real-time intelligence about customer perceptions, concerns, and emerging preferences at a scale and granularity that traditional survey-based research cannot match. Natural language processing systems monitor customer communications across channels (such things as support tickets, chat transcripts, review platforms, and social media) to identify sentiment trends, emerging complaints, and customer experience breakdowns as they develop rather than after they have become entrenched. This real-time intelligence enables customer experience teams to intervene before individual issues escalate and enables product and service teams to incorporate customer feedback signals into development priorities on a continuous basis.

The strategic value of *real-time customer intelligence* extends beyond individual issue resolution. Organizations that systematically connect customer sentiment data to operational and product decisions create feedback loops that continuously improve product-market fit, customer experience quality, and retention economics. This capability is particularly valuable in competitive markets where customer expectations are rising rapidly: organizations with real-time customer intelligence can anticipate and respond to shifts in customer expectations faster than competitors relying on periodic research.

AI has transformed marketing measurement and budget allocation by enabling more accurate attribution of revenue outcomes to specific marketing investments across the increasingly complex multi-channel customer journey. Traditional attribution models, such as last-touch, first-touch, or linear, provide simplistic and often misleading representations of how marketing investments combine to influence customer decisions. AI-driven multi-touch attribution and marketing mix modeling approaches use machine learning to estimate the *causal* contribution of each marketing channel and touchpoint to conversion outcomes, enabling more accurate budget allocation decisions that maximize revenue return on marketing investment.

Marketing mix optimization systems extend this capability by continuously recommending budget reallocations across channels, campaigns, and time periods based on predicted return on investment. These systems can incorporate constraints (e.g., regulatory requirements, brand safety considerations, channel capacity limits) and can model the lagged and interaction effects of marketing investments that human analysts typically cannot capture. Organizations that deploy AI-driven marketing attribution and optimization consistently demonstrate improvements in marketing efficiency and revenue impact that compound over time as the models incorporate additional data and become more accurate.

Operations, Supply Chain, and Logistics

Operations and supply chain management have been transformed by AI's ability to generate more accurate demand forecasts and to optimize inventory positioning decisions in real time across networks of suppliers, warehouses, distribution centers, and retail locations. Classical statistical forecasting methods (e.g., the Autoregressive Integrated Moving Average models, exponential smoothing, and their variants) perform adequately in stable environments but degrade rapidly when confronted with the volatility, structural breaks, and complex interaction effects that characterize modern supply chains. Machine learning forecasting models can incorporate a broader range of signals (including weather data, social media trends, economic indicators, competitive pricing, and promotional calendars) and can adapt more rapidly to structural changes in demand patterns.

The economic impact of improved demand forecasting compounds through the supply chain. More accurate demand signals reduce safety stock requirements, enabling lower inventory carrying costs without increasing stockout risk. Better forecasts improve supplier planning, reducing expediting costs and lead time variability. More precise demand estimates enable more efficient labor and capacity planning in distribution and retail operations. Organizations that have deployed advanced AI forecasting consistently report improvements in forecast accuracy of twenty to forty percent over statistical baselines, with corresponding improvements in inventory efficiency and service levels.

Transportation and logistics operations represent one of the most data-intensive and optimization-amenable domains in the enterprise. AI-powered routing and fleet management systems optimize delivery routes in real time across thousands of vehicles and hundreds of thousands of delivery points, incorporating real-time traffic data, weather conditions, vehicle telematics, time-window constraints, and dynamic load characteristics. The scale and computational complexity of these optimization problems far exceed what traditional operations research methods can address in real-time operational contexts; modern AI optimization systems using reinforcement learning and graph neural network architectures solve them continuously as conditions evolve.

As examples, UPS and DHL have pioneered large-scale deployment of AI-powered logistics optimization, with documented fuel savings and service reliability improvements that have delivered hundreds of millions of dollars in annual value. Beyond route optimization, AI contributes to fleet management through predictive maintenance systems that analyze telematics data to predict component failures before they cause vehicle downtime, enabling planned maintenance interventions that are dramatically less costly than roadside failures. Last-mile optimization (the most expensive and complex segment of the delivery value chain) is increasingly being addressed through AI systems that dynamically cluster deliveries, optimize sequence, and adapt routes in response to real-time events such as failed delivery attempts or customer rescheduling requests.

Predictive maintenance represents one of the highest-value and most broadly deployed AI applications in industrial operations. Traditional maintenance strategies (either time-based preventive maintenance or reactive breakdown maintenance) are economically suboptimal: time-based maintenance replaces components before they have reached the end of their useful life, wasting resources; reactive maintenance allows failures that are frequently more costly to address than planned interventions would have been. AI-powered predictive maintenance systems analyze sensor data from machinery and equipment to identify patterns that precede failure, enabling maintenance interventions to be scheduled at the optimal point in the asset lifecycle… after sufficient useful life has been extracted, but before failure occurs.

The value creation potential of predictive maintenance is substantial across multiple dimensions simultaneously. Reduced unplanned downtime directly increases production throughput and asset utilization. Extended component life through optimal intervention timing reduces materials costs. Planned maintenance interventions are less labor-intensive and disruptive than emergency repairs. Moreover, the data generated by predictive maintenance systems feeds back into equipment design and procurement decisions, enabling continuous improvement of asset reliability over successive equipment generations. Industrial operators in manufacturing, energy, transportation, and infrastructure have documented annual savings of tens of millions of dollars from AI-driven predictive maintenance programs, with payback periods typically measured in months rather than years.

AI-powered quality management systems are transforming defect detection and quality assurance processes across manufacturing, pharmaceutical production, food processing, and other industries where product quality is safety-critical or commercially significant. Computer vision systems trained on images of conforming and non-conforming products can identify defects with accuracy and consistency that human inspectors cannot match, particularly in high-speed production environments where the inspection window for each unit is measured in milliseconds. These systems operate continuously without the fatigue, distraction, or inconsistency that limit human inspection performance, and they can be trained to detect subtle defect signatures that are not reliably identifiable by the human eye.

Beyond end-of-line inspection, AI enables process control systems that monitor production parameters in real time and adjust process settings dynamically to maintain quality within specification, preventing defects from forming rather than detecting them after the fact. This shift from reactive to proactive quality management reduces scrap rates, rework costs, and the risk of quality escapes that reach customers… creating both direct cost savings and customer experience improvements that compound into competitive advantage.

The supply chain disruptions of 2020-2023, driven by the COVID-19 pandemic, geopolitical developments, extreme weather events, and semiconductor shortages demonstrated with painful clarity the economic costs of supply chain fragility and the limitations of traditional risk management approaches that relied on historical precedent and static scenario analysis. AI has emerged as a critical capability for supply chain risk management, enabling organizations to monitor a much broader range of risk signals in real time, model the propagation of disruptions through complex supply network graphs, and generate adaptive response recommendations under uncertainty.

AI-powered supply chain risk management systems ingest signals from diverse sources (e.g., supplier financial health indicators, geopolitical risk indices, weather forecasting models, port congestion data, and logistics capacity utilization) and use machine learning to generate early warnings of potential disruptions before they materialize in operational data. Network analysis tools model the multi-tier structure of supply networks, identifying hidden dependencies and single points of failure that traditional supply chain mapping cannot reveal. Simulation systems model the propagation of hypothetical disruptions across the network, enabling organizations to evaluate the resilience implications of sourcing decisions, inventory positioning, and supplier diversification strategies before disruptions occur.

Finance, Risk, and Compliance

Finance functions have historically been significant consumers of analytical talent performing tasks, such as data aggregation, variance analysis, report generation, and scenario modeling, that are amenable to AI augmentation and, in many cases, substantial automation. AI-powered financial planning and analysis platforms reduce the time required to close the books, generate management reports, and produce financial forecasts, freeing finance professionals to focus on the interpretive and strategic work that requires human judgment. Machine learning forecasting models applied to revenue, expense, and cash flow planning can incorporate a broader range of internal and external signals than traditional spreadsheet-based models, improving forecast accuracy and enabling more granular sensitivity analysis.

Dynamic financial forecasting, in which financial models are continuously updated with new data and regenerated on demand rather than refreshed on fixed reporting cycles, is becoming standard practice in AI-mature finance organizations. This shift from periodic to continuous forecasting enables faster management decision-making, more responsive resource allocation, and earlier identification of financial performance deviations that require intervention. The organizational implication is significant: finance teams in AI-mature organizations spend proportionally less time on data assembly and report production and more time on performance analysis, business partnership, and strategic planning… a role expansion that increases the function's strategic value to the enterprise.

Fraud detection represents one of the earliest and most mature applications of AI in financial services, and one of the highest-value deployments of machine learning in any functional domain. AI-powered fraud detection systems analyze transaction data in real time to identify patterns consistent with fraudulent activity (e.g., account takeover, payment fraud, synthetic identity fraud, money laundering) and generate risk scores that trigger review, step-up authentication, or transaction blocking within milliseconds of transaction initiation. The performance advantage of AI-based fraud detection over a rule-based system is substantial: machine learning models adapt continuously to evolving fraud patterns, detect subtle multivariate patterns that rule-based systems cannot express, and generate fewer false positives that inconvenience legitimate customers.

Financial institutions have documented fraud loss reductions of thirty to fifty percent following the deployment of advanced machine learning fraud detection systems, with simultaneous improvements in legitimate transaction approval rates that reduce customer friction and revenue loss from false declines. Anti-money laundering and financial crime detection systems apply similar machine learning approaches to the detection of complex multi-transaction patterns consistent with layering and integration… the stages of money laundering that individual transaction monitoring misses. These systems incorporate network analysis to identify suspicious relationships between accounts, entities, and transaction flows that alert human investigators to patterns warranting deeper examination.

AI has transformed credit risk assessment by enabling the incorporation of non-traditional data sources and more sophisticated modeling approaches than the credit scoring methodologies that dominated consumer and commercial lending for decades. Traditional credit scoring models rely primarily on credit bureau data such as payment history, outstanding balances, credit utilization, and are supplemented by application data. AI-powered underwriting models can incorporate alternative data sources (e.g., rent and utility payment history, banking transaction patterns, employment verification data, and in some applications digital behavioral signals) to generate more accurate risk assessments, particularly for borrowers with limited traditional credit histories.

The commercial implications of more accurate credit risk assessment compound across the credit lifecycle. At origination, better risk models enable more precise pricing, more appropriate credit limit assignment, and expansion of credit availability to creditworthy borrowers who would have been declined under traditional models. In portfolio management, AI-powered monitoring systems provide early warning of deteriorating credit quality in individual accounts and portfolio segments, enabling timely intervention before losses materialize. Stress testing and scenario analysis systems enable more granular assessment of portfolio resilience under adverse macroeconomic conditions, informing capital planning and risk appetite decisions.

Compliance functions face a dual challenge: the volume and complexity of regulatory requirements continue to expand across financial services, healthcare, manufacturing, and other regulated industries, while at the same time the organizational resources available to manage compliance obligations become more constrained. AI is addressing this challenge across multiple dimensions simultaneously. Natural language processing systems monitor regulatory publications (including legislation, regulatory guidance, enforcement actions, and industry standards) to identify changes relevant to the organization's compliance obligations, enabling faster regulatory change management. Compliance monitoring systems apply machine learning to transaction, communication, and operational data to identify potential regulatory violations before they become enforcement issues.

Regulatory reporting automation is another high-value AI application in compliance-intensive industries. AI systems that can extract structured data from unstructured documents, validate data against regulatory requirements, and generate formatted regulatory submissions reduce the time and manual effort required for regulatory reporting while improving accuracy and auditability. In financial services, AI-powered trade surveillance systems monitor trading activity for patterns consistent with market manipulation, insider trading, or best execution failures, thus enabling compliance teams to focus human review on the highest-risk alerts rather than reviewing large volumes of flagged items manually.

Human Resources and Talent Management

Talent acquisition is a domain in which AI has simultaneously delivered substantial efficiency improvements and raised significant ethical questions that organizations must address proactively. AI-powered recruitment systems automate resume screening, candidate-role matching, and initial engagement at a scale that manual processes cannot approach, reducing time-to-hire and enabling recruiting teams to evaluate a larger and more diverse candidate pool than resource constraints would otherwise permit. Resume parsing systems extract structured information from unstructured documents and apply matching algorithms that assess candidate qualifications against role requirements. Conversational AI systems handle initial candidate engagement, screening, and scheduling… thus replacing high-volume but low-complexity interactions that consumed substantial recruiter time.

The ethical dimension of AI in talent acquisition is serious and requires explicit organizational attention. AI systems trained on historical hiring data can perpetuate and amplify historical biases embedded in that data that systematically disadvantages candidates from demographic groups that were underrepresented in historical hires, or encoding the preferences of specific hiring managers in ways that introduce subjective bias into ostensibly objective selection processes. Organizations deploying AI in recruitment must invest in algorithmic bias auditing, diverse training data curation, fairness constraint implementation, and ongoing monitoring of demographic outcomes. Transparency with candidates about the role of AI in selection processes is both an ethical obligation and, in many jurisdictions, an emerging regulatory requirement.

Employee attrition is one of the most economically significant and predictable risks that organizations face, and it is a domain in which AI-powered predictive analytics has demonstrated consistent value. *Attrition prediction models* analyze a combination of employee tenure, performance trajectory, compensation position relative to market, engagement survey responses, manager relationship indicators, workload metrics, and career development signals to generate individual-level attrition risk scores that enable proactive retention interventions. Organizations that deploy these systems report meaningful reductions in voluntary attrition rates among high-value employee populations, with corresponding reductions in replacement hiring costs that are typically several times annual salary for professional and technical roles.

Workforce planning systems extend predictive analytics to the aggregate level, modeling future workforce supply and demand under various business scenarios and identifying skill gaps that require targeted hiring, development, or redeployment. These systems integrate internal HR data with external labor market data (including skills availability, compensation benchmarks, and talent migration patterns) to inform workforce strategy decisions with market intelligence that manual research cannot provide at comparable depth or currency. The integration of workforce planning with financial planning systems enables organizations to model the workforce cost implications of business strategy decisions and to optimize talent investment allocation across competing priorities.

Learning and development activities are undergoing a transformation driven by AI's ability to personalize skill development content and sequencing at the individual level, and to connect learning investment to measurable business performance outcomes. Traditional corporate learning programs deliver standardized content to cohorts of employees at fixed intervals… an approach that is efficient to administer but poorly aligned with individual learning needs, career trajectories, and organizational skill gaps. AI-powered learning platforms analyze individual skill profiles, role requirements, career aspirations, and learning style preferences to generate personalized learning pathways that deliver the right content, in the right format, at the right moment in the learning journey.

Large language models have substantially expanded the capability of AI learning systems in this area. These model-powered learning platforms can generate customized explanations, examples, and practice scenarios adapted to individual knowledge levels, and can engage in Socratic dialogue that promotes deeper understanding than passive content consumption. AI tutoring systems that provide immediate, personalized feedback on learning activities have demonstrated significantly better skill acquisition outcomes than self-paced content delivery alone, particularly for complex technical skills that require practice and feedback to develop. Skills intelligence platforms that map individual skill profiles across the workforce enable organizations to identify skill gaps, target development investment, and make more informed decisions about internal mobility, team composition, and succession planning.

AI is expanding the HR function's capacity to understand and improve employee experience in real time, moving beyond annual engagement surveys to continuous listening architectures that capture employee sentiment and experience signals on an ongoing basis. *Natural language processing systems* analyze the content of pulse surveys, open-text feedback channels, and anonymous communication platforms to identify emerging themes, concerns, and sentiment shifts that enable HR and management to intervene before issues become entrenched. The analytical sophistication of these systems (capable of distinguishing between isolated complaints and systemic issues, and of identifying the organizational locations, roles, or management relationships most strongly associated with negative experience) enables more targeted and effective HR interventions than survey-based analysis alone.

AI is also expanding HR's ability to support employee wellbeing through personalized benefits guidance, mental health resources, and workplace accommodation recommendations. Conversational AI systems that provide employees with personalized guidance on benefits selection, wellness programs, and support resources reduce the friction associated with accessing HR services, while generating behavioral data that enables HR to identify populations with elevated wellbeing needs and target support resources accordingly. Organizations should approach employee data collection and analysis in this sensitive domain with particular care for privacy, consent, and the ethical implications of using behavioral data to infer mental health status… areas in which organizational policies must anticipate evolving regulatory requirements.

Research, Innovation, and Product Development

Research and development (R&D) has been transformed by AI's ability to accelerate the generation and evaluation of hypotheses across scientific domains, compressing the timelines and reducing the cost of discovery processes that traditionally required years of experimental iteration. In pharmaceutical research, AI models trained on molecular structure databases can predict the binding affinity, toxicity, and bioavailability properties of candidate molecules, enabling researchers to prioritize experimental synthesis and testing on the candidates most likely to exhibit favorable properties. This capability has compressed the early-stage drug discovery timeline from years to months for specific target classes, fundamentally altering the economics of pharmaceutical R&D.

DeepMind's AlphaFold system, which achieved breakthrough accuracy in protein structure prediction, illustrates the potential of AI to resolve longstanding scientific problems with implications that extend across biology, materials science, and medicine. The strategic implications for life science organizations are profound: organizations that develop or access AI-powered structural biology capabilities can dramatically accelerate target identification, mechanism understanding, and drug candidate design relative to competitors relying on experimental methods alone. Similar AI-driven acceleration is occurring across materials science, chemistry, and climate technology, where machine learning surrogate models enable rapid exploration of design spaces that experimental methods cannot traverse at comparable speed.

Product development functions benefit from AI across the full development lifecycle… essentially from market opportunity identification through concept generation, design optimization, testing, and validation. AI-powered market intelligence systems synthesize data from diverse sources (e.g., customer feedback, competitive analysis, patent filings, technical literature, and social media trends) to identify market opportunities and emerging customer needs before they become widely apparent. This intelligence enables organizations to initiate development programs aligned with future market conditions rather than current ones, improving the probability of market success.

Generative design tools use AI to explore design alternatives within specified constraint sets, namely material properties, manufacturing processes, cost targets, and performance requirements, generating solution candidates that human designers would not have considered and that often achieve superior performance relative to conventionally derived designs. Simulation-driven development, in which AI-powered digital twins enable virtual evaluation of product performance under a wide range of operating conditions, reduces the number of physical prototype iterations required and compresses development timelines while improving the quality of designs that enter production. Engineering organizations that have deployed AI-powered simulation consistently report meaningful reductions in time-to-market and development cost for comparable product categories.

AI has significantly expanded organizational capability for intellectual property analysis and competitive technology intelligence. Natural Language Processing systems can analyze patent databases comprising millions of documents to identify relevant prior art, assess the novelty of proposed innovations, map the patent landscape around specific technology domains, and identify white space opportunities where intellectual property protection may be available. This capability, which previously required specialist patent attorneys and substantial manual research time, can now be accessed by engineering and product development teams through AI-powered tools that return results in minutes rather than weeks.

Innovation portfolio management or the organizational discipline of tracking, evaluating, and prioritizing a pipeline of development initiatives benefits from AI systems that can assess the technical maturity, commercial potential, and competitive landscape of individual initiatives more systematically than subjective management review processes. Organizations that apply AI to innovation portfolio management report improvements in the quality of portfolio decisions, more consistent application of evaluation criteria across initiatives, and better alignment of development investment with strategic priorities.

Information Technology and Cybersecurity

Information technology operations have been transformed by AI's ability to process the massive volumes of telemetry data generated by modern IT infrastructure (e.g., server metrics, application logs, network traffic, container orchestration events) and to identify patterns that indicate impending performance degradation, security threats, or system failures before they cause service disruptions. AIOps platforms apply machine learning to this telemetry data to automate anomaly detection, correlate events across infrastructure layers to identify root causes, and generate remediation recommendations that reduce mean time to resolution for incidents.

The operational value of AIOps is substantial in environments where manual monitoring of telemetry data at scale is impossible and where service reliability has direct revenue implications. Organizations operating cloud-native applications, microservices architectures, or high-volume transactional systems have documented significant reductions in incident frequency, incident duration, and the labor cost of IT operations management following AIOps deployment. Automated remediation capabilities (particularly those which allow AIOps systems to execute pre-approved remediation actions without human intervention for well-understood failure patterns) further reduce response time and free operations teams for higher-value analytical and architectural work.

Cybersecurity represents a domain in which the deployment of AI by defenders is a strategic necessity, because adversaries are already deploying AI to enhance the sophistication, scale, and speed of cyberattacks. AI-powered security operations center (SOC) systems analyze security events across endpoint, network, cloud, and application layers to detect threats that traditional signature-based detection systems miss… including novel attack variants, living-off-the-land techniques, and advanced persistent threat behaviors that evolve to evade static detection rules. Machine learning systems trained on historical attack data can identify behavioral anomalies (e.g., unusual authentication patterns, abnormal data exfiltration volumes, or unexpected lateral movement) that indicate compromise even when specific malware signatures are absent.

Extended Detection and Response (XDR) platforms represent the *current state of the art* in AI-powered security operations, correlating security signals across multiple control planes to provide unified threat visibility and automated response capabilities that exceed the capacity of human analysts working with siloed security tools. Threat intelligence platforms use AI to analyze threat data from diverse sources, such as security research publications, dark web forums, honeypot systems, or industry sharing networks, to generate actionable intelligence about emerging threats and enable proactive defensive hardening before specific attack variants materialize in the organization's environment. AI is also being applied to vulnerability management, prioritizing the remediation of security vulnerabilities based on predicted exploitability and organizational exposure rather than treating all vulnerabilities as equivalently urgent.

Software development itself has been transformed by AI-powered coding assistance tools that accelerate developer productivity, improve code quality, and reduce the burden of routine development tasks. AI coding assistants (including GitHub Copilot, Cursor, and similar tools) generate code completions, function implementations, and test cases from natural language descriptions and contextual cues, enabling developers to implement functionality at significantly higher velocity than manual coding allows. Studies of developer productivity following the adoption of AI coding assistants consistently report task completion time reductions of thirty to fifty percent for common development activities, with particularly large gains for boilerplate code generation, documentation, and test writing.

AI is also being deployed to improve code quality, security, and reliability through automated code review systems that identify bugs, security vulnerabilities, and style violations with greater coverage and consistency than manual review. AI-powered test generation systems create comprehensive test suites from code analysis, covering edge cases and boundary conditions that human test writers frequently miss. Architectural analysis tools use AI to identify technical debt patterns, dependency risks, and performance bottlenecks in large codebases that are too complex for human architects to comprehend in their entirety. Together, these capabilities are shifting the composition of developer work from code writing toward higher-level system design, product definition, and AI supervision… a transformation with significant implications for software development workforce planning.

Legal, Regulatory Affairs, and Corporate Governance

Legal functions have historically been resource-constrained relative to demand, with high-volume but structurally repetitive tasks (e.g., contract review, due diligence, or legal research) consuming attorney time that could more productively be devoted to complex advisory and advocacy work. AI is addressing this structural imbalance by automating or substantially augmenting the most repetitive legal tasks, enabling legal teams to handle larger volumes of work at lower cost while improving consistency and reducing the risk of human error in high-volume document review processes.

AI-powered contract analysis systems can review and extract structured information from thousands of contracts in the time it would take human reviewers to examine dozens, identifying non-standard provisions, unusual risk allocations, and compliance violations across large contract portfolios. These systems have proven particularly valuable in mergers and acquisitions (M&A) due diligence, where the volume and time pressure of contract review has historically been a significant constraint on deal execution. AI-powered contract lifecycle management platforms extend these capabilities to contract creation… generating draft agreements from structured inputs, incorporating organizational playbook provisions, and flagging deviations from standard terms for attorney review.

Organizations operating in complex regulatory environments face the challenge of monitoring an expanding volume of regulatory activity across multiple jurisdictions and incorporating regulatory changes into operational practice on compressed timelines. AI-powered regulatory intelligence systems address this challenge by continuously monitoring regulatory sources such as legislative databases, agency publications, court decisions, enforcement actions, and international standards, to identify developments relevant to the organization's regulatory obligations.

Natural language processing systems classify regulatory developments by relevance, jurisdiction, and operational impact, enabling compliance and regulatory affairs teams to prioritize their review attention and accelerate the regulatory change management process. In pharmaceuticals and medical devices, AI systems that maintain structured knowledge bases of regulatory requirements and map those requirements to product characteristics and clinical data enable more efficient regulatory submission preparation and more responsive management of regulator inquiries. In financial services, regulatory horizon scanning systems that track supervisory guidance and enforcement trends enable proactive compliance posture adjustment before regulatory examination or enforcement action.

Cross-Functional AI Integration: The Enterprise Intelligence Architecture

The value created by AI deployment at the individual functional level is significant, but it is substantially amplified when AI systems across functions are *architecturally* integrated... sharing data, exchanging insights, and coordinating decisions in ways that reflect the genuine interdependencies of enterprise operations. A demand forecast that informs only inventory planning delivers a fraction of the value of a demand forecast that simultaneously informs inventory positioning, marketing campaign timing, workforce scheduling, supplier purchase orders, and financial revenue planning. The incremental cost of cross-functional integration is modest relative to the incremental value it creates, yet many organizations treat functional AI deployment as an end unto itself rather than as the foundation for *enterprise intelligence architecture.*

Cross-functional AI integration creates value through several distinct mechanisms. Data sharing allows AI to use information from multiple functions, such as marketing, finance, and operations, to build a more accurate customer model than relying on just one data source. Insight propagation enables predictions generated by AI systems in one function to inform decisions in others and thus allowing a predictive churn signal from the customer analytics system to trigger proactive account management actions in sales before the customer contacts the service team. Decision coordination enables AI systems across functions to optimize decisions jointly rather than locally... preventing the suboptimal outcomes that arise when individually rational functional decisions are collectively inefficient.

Achieving cross-functional AI integration requires deliberate architectural investment in the data infrastructure, API design, and governance frameworks that enable AI systems across functions to communicate and share state. As noted previously, several architectural patterns have emerged as effective approaches to this challenge. The enterprise data platform model centralizes data from across functional systems in a unified data warehouse or data lakehouse, providing a common substrate from which functional AI systems can access cross-functional signals and into which they can publish their outputs. This model simplifies data governance and enables ad hoc cross-functional analysis but requires careful design to avoid becoming a governance bottleneck.

The event-driven architecture model connects functional systems through a shared event streaming platform that broadcasts significant events (e.g., transactions, customer interactions, inventory movements, or production outputs) to subscribing systems in real time. AI systems that consume such event streams can incorporate cross-functional signals without tight coupling to the originating systems, enabling more flexible and scalable integration than point-to-point data sharing. The AI-mediated orchestration model uses an AI orchestration layer that coordinates decisions across functional AI systems... explicitly modeling the interdependencies between functional decisions and optimizing joint outcomes rather than optimizing each functional decision independently. This model is architecturally complex but enables the *highest* level of cross-functional decision optimization.

A concrete illustration of cross-functional integration value: a consumer goods organization with integrated AI systems across demand forecasting, supply chain planning, marketing optimization, and financial planning can execute the following sequence automatically. The demand forecasting system detects an emerging demand signal for a specific product category based on social media trends and early transaction data. The supply chain planning system receives this signal and automatically identifies the inventory reposition actions and supplier orders required to support projected demand. The marketing optimization system incorporates the demand signal to schedule promotional investment for the affected category in advance of the demand peak. The financial planning system updates revenue, cost, and cash flow projections based on the integrated operational plan. This sequence, which would require weeks of manual coordination across functional teams operating with siloed data systems, executes in hours in an integrated AI architecture... with compounding improvements in revenue capture, working capital efficiency, and forecast accuracy.

Data Foundations for Functional AI Transformation

The effectiveness of AI systems across all functional domains is fundamentally constrained by the quality of the data on which they are trained and operated. This is not a novel observation (the aphorism 'garbage in, garbage out' predates AI by decades) but it is one that organizations consistently underestimate when planning AI deployments, with the result that model performance in production frequently falls short of development-environment benchmarks because production data quality is lower than training data quality. Data quality for AI purposes encompasses multiple dimensions: accuracy (does the data correctly represent the phenomenon it is intended to capture?), completeness (does the data cover the full range of cases relevant to the model?), consistency (is the same entity or event represented consistently across data sources and time periods?), timeliness (is the data sufficiently current to represent current patterns?), and granularity (does the data capture the level of detail required for the intended AI application?).

Investing in data quality as a prerequisite for AI deployment, rather than treating data quality improvement as a parallel or post-hoc activity, is one of the most important strategic choices in functional AI transformation. Organizations that have made systematic investments in data governance, data quality management, and master data management before deploying functional AI consistently achieve better model performance, faster time-to-value, and lower ongoing maintenance costs than those that deploy AI on poorly governed data and then attempt to remediate quality issues after deployment.

Data quality is a necessary but not sufficient condition for effective functional AI deployment. Data accessibility (i.e., the ability of AI systems to access relevant data from across the enterprise on the timelines required for their use cases) is an equally important architectural requirement that many organizations fail to address adequately. AI systems that require real-time inference need access to data in milliseconds; batch training systems need access to historical data spanning months or years; monitoring systems need access to streaming data that reflects current operational conditions. Meeting these diverse accessibility requirements demands an intentional data platform architecture that supports multiple data consumption patterns simultaneously.

The dominant enterprise data platform architectures of the current period (i.e., cloud data warehouses, data lakehouse architectures, and streaming data platforms) each address different subsets of AI data accessibility requirements. Organizations that standardize on a single architecture often find it inadequately suited to some of their AI use cases; organizations that deploy multiple architectures face data integration complexity and governance challenges. Designing an enterprise data platform architecture that meets the data accessibility requirements of the full portfolio of planned AI applications, rather than the requirements of the initial deployment use cases, is a critical strategic investment that pays dividends across the full AI transformation timeline.

Data governance for AI encompasses both the traditional data governance concerns (i.e., data ownership, access control, quality standards, and lifecycle management) and AI-specific governance requirements that traditional data governance frameworks do not fully address. AI-specific data governance concerns include the documentation of data lineage from source to model training to production inference, which is required for model auditability and regulatory compliance; the management of training data quality and bias characteristics, which determine model fairness and reliability; the governance of data used for model evaluation and monitoring, which must be representative of production data distributions to provide reliable performance signals; and the management of sensitive personal data used in AI systems, which carries privacy and regulatory compliance obligations that extend beyond those applicable to data used in traditional analytics.

Organizations that develop comprehensive data governance frameworks for AI (addressing both traditional and AI-specific requirements) create a governance foundation that accelerates deployment velocity, reduces regulatory risk, and enables more confident organizational adoption of AI-generated insights. Governance frameworks that are overly restrictive, however, can impede the data sharing and experimentation that effective AI development requires. Calibrating governance requirements to the sensitivity of the data and the risk profile of the intended AI application (applying stricter governance to sensitive personal data and high-stakes decision applications, and lighter governance to internally generated operational data and low-stakes applications) balances the protection of legitimate interests with the facilitation of AI innovation.

Talent, Culture, and Change Management for Functional AI Transformation

The most consequential determinant of functional AI transformation success is not the sophistication of the AI technology deployed but the organization's ability to integrate AI capabilities into *human workflows* in ways that generate genuine performance improvements. This integration challenge is fundamentally a human and organizational challenge: it requires employees to develop new skills, managers to redesign workflows and decision processes, and organizational cultures to shift from intuition-based to data-informed decision-making norms. Technology implementations that do not address these human dimensions consistently underperform relative to their technical potential.

Organizational change management for functional AI transformation begins with a clear articulation of how AI will change individual roles and responsibilities. Employees are more willing to engage with AI transformation when they understand specifically how AI tools will affect their work, what new skills they will need to develop, and what the organization's commitment will be in supporting them through the transition. Ambiguity about AI's role in the workforce fuels resistance and disengagement that undermines adoption; clarity and transparency (even when the news is that some tasks will be automated) enables constructive adaptation.

Effective functional AI transformation requires AI literacy across the full organization… not only in the technical teams that build and maintain AI systems, but in the business professionals who work alongside AI tools, the managers who make decisions informed by AI outputs, and the executives who set the strategic direction for AI investment. AI literacy for non-technical professionals encompasses the ability to understand the capabilities and limitations of AI systems, to critically evaluate AI-generated outputs, to identify when AI recommendations should be followed and when human judgment should override them, and to participate meaningfully in the governance and accountability processes that determine how AI is used.

Business organizations that invest in building AI literacy across the workforce (through formal training programs, experiential learning opportunities, and leadership modeling of data-informed decision-making) demonstrate consistently better functional AI adoption rates and business outcomes than those that treat AI literacy as a specialized technical competence. The investment required is not trivial: meaningful AI literacy development across a large organization requires sustained curriculum development, delivery infrastructure, and management reinforcement. But the return on this investment, in terms of adoption velocity, model utilization quality, and organizational capacity to identify and exploit new AI opportunities, is substantial.

Effective functional AI transformation typically requires explicit redesign of work processes and decision rights to reflect the changed role of human judgment in an AI-augmented environment. When AI systems take over routine decision-making tasks (such as transaction approvals, scheduling decisions, or inventory reorder recommendations) human roles shift from executing those decisions to supervising AI performance, handling exceptions, and making the complex, context-sensitive judgments that AI systems cannot reliably produce. This shift requires explicit redefinition of role expectations, performance metrics, and management accountability structures.

Decision rights (meaning the formal and informal assignment of who makes which decisions and under what conditions) are a particularly important design element in AI-augmented organizations. AI systems that generate recommendations without clear governance about when those recommendations should be followed, when they should be overridden, and what escalation processes apply in ambiguous cases tend to produce inconsistent outcomes: sometimes AI is ignored when it should be followed, and sometimes AI recommendations are followed when human judgment should override them. Explicit decision rights frameworks that specify the conditions under which AI recommendations are authoritative versus advisory, and that assign clear accountability for outcomes to humans who retain decision authority, create the governance structure needed for AI to generate consistent value while preserving appropriate human oversight.

Measuring and Managing Functional AI Performance

Measuring the impact of functional AI transformation requires a measurement architecture that spans multiple levels, from technical model performance through process efficiency to business outcomes, and that maintains clear linkages between each level. The temptation to evaluate AI systems primarily on technical metrics (e.g., model accuracy, precision, recall, or latency) is understandable given the availability of these metrics and the ease of computing them, but it is strategically inadequate. Technical metrics are proxies for business impact, not measures of it, and the relationship between technical performance and business value is often non-linear and "context-dependent."

A functional AI measurement framework should articulate, for each AI application, the chain of causation from model outputs through process changes to business outcomes: what decisions or actions does the model inform, how do those decisions affect process performance, and how does process performance translate into business outcomes? This causal chain provides the foundation for attribution (i.e., connecting observed business outcome changes to AI deployment) and for identifying where in the chain performance improvements are needed when business outcomes fall short of targets.

Each functional domain requires a set of KPIs calibrated to the specific value creation mechanisms of AI in that domain. Marketing AI performance should be assessed against metrics including customer acquisition cost, conversion rate improvement, customer lifetime value, and marketing return on investment... not simply against engagement metrics that may not translate into revenue. Sales AI performance metrics include pipeline velocity, win rate, forecast accuracy, and sales productivity per representative. Operations AI metrics include forecast accuracy improvement, inventory turns, on-time delivery rates, maintenance cost per asset, and quality defect rates. Finance AI metrics include fraud loss rates, false positive rates, model accuracy for credit risk, and compliance finding rates. HR AI metrics include time-to-hire, retention rates for targeted employee populations, training completion and skill acquisition rates, and workforce cost per unit of output.

Across all functional domains, AI governance metrics (e.g., model fairness indicators, data quality scores, model drift measures, and bias audit outcomes) should be tracked alongside business performance metrics to ensure that AI systems are *not* achieving business results through means that create regulatory, reputational, or ethical risks. These governance metrics should be reported to the same leadership audiences that review business performance metrics, not treated as technical concerns managed exclusively within the AI engineering team.

AI systems deployed in functional operations are not static artifacts; they are dynamic systems that require ongoing maintenance, monitoring, and improvement to maintain their performance as the environments in which they operate evolve. Model drift, the degradation of model performance that occurs as the statistical distribution of input data diverges from the distribution on which the model was trained, is an inevitable phenomenon that requires systematic monitoring and periodic model retraining or replacement to manage. Organizations that treat AI deployment as a project with a defined end date, rather than as an ongoing operational commitment, consistently find that model performance degrades over time and that the business value of AI investments diminishes accordingly.

Continuous improvement frameworks for functional AI systems should define the monitoring metrics and thresholds that trigger model review, the processes for investigating and diagnosing performance degradation, and the decision criteria for model retraining versus model replacement. They should also define the organizational responsibilities for model monitoring… a governance gap in many organizations, where the team that built the model is no longer responsible for its performance after deployment, but the operational team that uses it lacks the technical capability to monitor or improve it. Assigning clear, sustained accountability for model performance to specific organizational roles is a prerequisite for maintaining the value of functional AI systems over their operational lifecycle.

Responsible AI in Functional Deployment

Functional AI applications that influence consequential decisions (e.g., hiring, credit, medical diagnosis, performance evaluation, or law enforcement) carry significant risks of producing outcomes that are unfair to individuals from protected demographic groups, particularly when AI systems are trained on historical data that reflects past discrimination or structural inequality. Organizations deploying AI in these high-stakes functional contexts have both ethical obligations and, in many jurisdictions, legal compliance requirements to assess, manage, and monitor model fairness.

Algorithmic fairness assessment requires organizations to define the relevant fairness criteria for each application (which may include demographic parity, equalized odds, individual fairness, or other formulations) and to evaluate model outputs against those criteria across protected demographic groups. Where fairness violations are identified, organizations must decide whether to mitigate them through data curation, model retraining, post-hoc adjustment of model outputs, or process redesign… each of which involves trade-offs with model accuracy, operational efficiency, or other values. The mitigation strategy appropriate for a specific application depends on the legal requirements, the magnitude of the fairness violation, and the operational constraints of the deployment context.

Many functional AI applications make or inform decisions that affect individuals who have a legitimate interest in understanding why a decision was made and how it can be challenged. Credit decisions, hiring decisions, medical diagnoses, and performance evaluations are among the domains where regulatory requirements for explanation exist or are emerging. The degree of explainability required varies by context: a customer who receives a personalized product recommendation has a lower explainability expectation than a loan applicant who is declined, whose employment application is rejected, or whose insurance claim is denied.

Explainable AI (XAI) techniques (including feature importance methods, counterfactual explanations, and model-agnostic interpretation approaches) have advanced substantially, enabling organizations to generate meaningful explanations for AI-informed decisions in many contexts without sacrificing model performance. Organizations should invest in *explainability infrastructure* as part of their functional AI deployment architecture, not as an afterthought when regulatory or customer demands require it. Proactive investment in explainability supports regulatory compliance, builds customer and employee trust, and enables more effective human oversight of AI systems… all of which contribute to the durability of AI value creation.

As XAI systems become more deeply embedded in functional decision workflows, maintaining meaningful human oversight becomes both more important and more organizationally challenging. The risk of over-reliance on AI (i.e., deferring to AI recommendations without applying the critical judgment that human oversight is intended to provide) is a documented *failure mode* in a wide range of AI applications, from radiology AI to algorithmic trading to content moderation. Organizations must design *human* oversight mechanisms that are genuinely effective, not merely nominal: oversight processes in which human reviewers lack the time, information, or skills to critically evaluate AI recommendations are not meaningfully different from fully automated decision processes from an accountability and error-correction perspective.

In many functional contexts, preserving the right to human review for individuals affected by AI-informed decisions is both an ethical requirement and, under regulations including the EU AI Act and GDPR, a legal one. Organizations should design functional AI workflows with explicit provisions for human escalation, independent review of AI recommendations, and correction of AI errors… and should invest in the training and tools required for human reviewers to exercise these functions meaningfully. The organizational culture around AI oversight is also important: cultures that treat AI recommendations as authoritative absent compelling contrary evidence are more vulnerable to systematic AI errors than cultures that treat AI as a powerful input to human judgment rather than a replacement for it.

Implementation Roadmap: From Functional Pilot to Enterprise Transformation

Organizations beginning or accelerating functional AI transformation face the challenge of sequencing investments across multiple functions and use cases in ways that maximize early value creation, build the organizational capabilities required for more advanced deployments, and avoid overextending limited implementation capacity. A well-designed implementation roadmap reflects both the strategic priorities of the organization and the practical sequencing logic of AI deployment: foundational data infrastructure must precede advanced modeling; organizational AI literacy must develop in parallel with technical deployment; governance frameworks must be established before high-risk applications are deployed.

A useful heuristic for sequencing functional AI investment prioritizes use cases that combine high business value with data readiness, organizational receptivity, and implementation feasibility. Early deployments that deliver visible value quickly build organizational confidence in AI and create the political capital required for larger and more complex investments. Use cases that require extensive data infrastructure development, significant change management, or novel regulatory navigation are better positioned in later roadmap phases, after foundational capabilities have been established and organizational confidence has been built.

The organizational model for delivering functional AI transformation involves a fundamental design choice between centralized and distributed AI capability. Centralized AI Centers of Excellence concentrate data science, ML engineering, and AI governance expertise in a shared services function that works with functional teams on a project basis… an approach that enables knowledge sharing, consistent practices, and efficient use of scarce technical talent. Distributed or embedded AI team models place technical capability directly within functional teams, enabling closer integration with domain expertise and business context but risking fragmentation of technical practices and duplication of foundational infrastructure.

As noted previously, leading organizations have generally evolved toward hybrid models that combine a central platform team (responsible for data infrastructure, ML platform, governance frameworks, and technical standards) with embedded capability within functional domains that combines technical AI skills with deep domain expertise. This *hybrid* model requires careful design of the interface between central and functional teams, including clear delineation of responsibilities, service level agreements for platform support, and governance mechanisms that ensure consistent standards across the distributed deployment landscape.

Sustaining executive and board-level commitment to functional AI transformation over the multi-year timelines required for meaningful impact demands a measurement and communication framework that tracks and reports transformation progress in terms that resonate with strategic and financial stakeholders. Progress metrics should include not only the operational and financial impacts of individual AI deployments but also the organizational capabilities (e.g., data infrastructure maturity, AI talent depth, governance framework development, and cultural readiness indicators) that determine the organization's trajectory and ultimate transformation potential. Transformation narratives that report only on current deployment outcomes without describing the capability foundations being built for future value creation can create unrealistic expectations and undermine long-term investment commitment.

AI as the Operating System of the Intelligent Enterprise

AI-powered functional transformation is not a set of technology deployments to be completed and checked off a project list. It is an ongoing organizational capability that, when developed systematically across functional domains and integrated into a coherent enterprise intelligence architecture, becomes the operating system of the intelligent enterprise... continuously learning from operational experience, anticipating emerging challenges, optimizing decisions across the full scope of organizational activity, and adapting to changing conditions faster than competitors operating without AI-native capabilities.

The organizations that achieve this state have invested in each of the dimensions examined in this chapter: the data foundations that enable AI systems to learn from experience; the technical infrastructure that enables AI systems to operate at production scale; the talent and culture that enable human-AI collaboration to generate value that neither human nor machine intelligence could produce alone; the governance frameworks that ensure AI systems operate within ethical and regulatory bounds; the measurement architectures that connect AI performance to business outcomes; and the cross-functional integration that multiplies the value of functional AI investments.

The distance between where most organizations are today and the *intelligent enterprise* of this description is substantial... but the path is well-mapped, the tools are increasingly accessible, and the competitive consequences of the journey are clear. Organizations that develop systematic, sustained, and strategically coherent AI transformation programs across their functional operations will find that the compounding benefits of AI-driven performance improvement create advantages that grow more durable with each passing year. The intelligent enterprise is not a destination to be reached once and maintained effortlessly; it is a capability that is built, deepened, and extended continuously… and that rewards sustained commitment with compounding competitive advantage.

CHAPTER 10

Responsible AI, Governance, and Regulation

AI Ethics, Fairness, and Bias Mitigation

AS ARTIFICIAL INTELLIGENCE BECOMES INTEGRAL to business operations, organizations must confront not only technical challenges but also ethical, regulatory, and societal responsibilities. Responsible AI is no longer *optional*; it is *essential* for sustaining trust, mitigating risk, and enabling long-term regulatory compliance. Embedding ethical principles and governance structures into AI development and deployment allows organizations to balance innovation with accountability, fairness, and transparency… and, as the sections that follow will argue, to treat that balance as a *strategic capability* rather than a defensive cost.

AI ethics frameworks provide the foundation for responsible deployment. These frameworks establish guiding principles (e.g., respect for human rights, transparency, accountability, and the promotion of societal well-being) and organizations frequently adapt internationally recognized guidelines, such as those issued by the OECD, IEEE, or the European Commission, into internal policies and review processes. Ethics frameworks are most useful when they function not as static value statements but as *operational decision criteria*: they should guide concrete trade-offs during model design, data selection, deployment, and ongoing use, providing a structured basis for decision-making at every stage of the AI lifecycle rather than a document that is drafted once and consulted rarely.

Explainability, fairness, and bias mitigation are the practical mechanisms through which these principles are enforced. Explainable AI (XAI) enables humans to understand the reasoning behind model outputs, which supports both organizational legitimacy and regulatory compliance; as models grow more complex, the capacity to interrogate a decision rather than simply accept it becomes a condition of meaningful human oversight. Fairness requires identifying and minimizing disparate impacts across demographic groups, but the technical reality is more complicated than the principle suggests: several mathematically valid formulations of fairness (e.g., demographic parity, equalized odds, predictive parity) are frequently in tension with one another, and the appropriate criterion depends on the specific decision context and the values of the affected population. Governance frameworks should specify, for each application, which fairness criteria apply, how they are measured, and what remediation process is triggered when a violation is detected… a specificity that many organizational ethics statements lack. Bias mitigation, meanwhile, is best understood not as a one-time corrective step but as a continuing management responsibility, requiring ongoing attention to data sources, modeling choices, and real-world outcomes. Techniques such as re-weighting training datasets, auditing model predictions, and adversarial testing all reduce (but do not eliminate) the likelihood of systematic discrimination.

Much of this risk originates upstream, in the data itself. *Historical bias* reflects inequities already present in the world and thus embedded in any dataset that faithfully records it; *representational bias* arises when certain groups are numerically underrepresented in the training population, degrading model performance for those groups regardless of the model's aggregate accuracy. A hiring model trained predominantly on résumés from a single geographic region, for instance, may systematically disadvantage qualified candidates from other regions… not because the model was designed to discriminate, but because its statistical picture of a "qualified candidate" was built from an unrepresentative sample. Because these biases can emerge or intensify after deployment, as input distributions shift and as models are repurposed for uses their designers did not anticipate, responsible organizations pair initial validation with continuous monitoring and feedback loops that surface *emerging* disparities as conditions change.

The regulatory frameworks that formalize these ethical obligations (most consequentially, the European Union's AI Act) are discussed in depth later in this chapter. What should be emphasized here is the relationship between the two: ethics frameworks are the substantive commitments an organization makes about how its AI systems should behave, and regulation is the external mechanism that increasingly requires organizations to demonstrate that those commitments are being met. Neither one substitutes for the other. An organization that only pursues regulatory minimums will find itself perpetually behind an accelerating compliance frontier; an organization that treats ethics as aspirational, without the audit trails and technical controls regulation demands, will find its commitments unenforceable and, ultimately, *unbelievable* to regulators, customers, and employees alike.

Intellectual Property and Data Sovereignty: Ownership, Output, and the Risks of Data Leakage in the Age of Generative AI

The arrival of generative AI (GenAI) in the business enterprise has created a legal and governance landscape that is, by any honest assessment, still being constructed in real time. Courts are issuing rulings that narrow (but do not close) the major open questions; regulatory bodies are publishing guidance that is frequently outpaced by technological capability; and the contractual frameworks that govern enterprise software relationships were almost universally written before the specific intellectual property questions raised by GenAI were legally or technically conceivable. Into this environment, business leaders and legal counsel are asked to make consequential decisions: what data to expose to which AI systems, how to assert ownership over AI-generated work product, how to protect proprietary information from inadvertent disclosure, and how to comply with data sovereignty requirements that vary dramatically across the jurisdictions in which global enterprises operate.

The difficulty is compounded because the costs of getting these decisions wrong are asymmetric and often invisible until they are not: a company that unknowingly trains a public model on its proprietary customer data, or that loses a copyright dispute because it cannot demonstrate sufficient human authorship in an AI-generated work, or that violates a data-residency requirement by routing sensitive information through cloud infrastructure located in the wrong jurisdiction, may not discover the exposure until it manifests as litigation, regulatory action, or competitive harm exceeding whatever efficiency gain the AI deployment was designed to produce. Understanding the intellectual property and data sovereignty dimensions of AI governance is not, in this environment, a compliance exercise to be delegated to counsel and forgotten; rather, it is a strategic risk management imperative that belongs at the center of every enterprise AI governance framework.

Who Owns the Output?

The question of who *owns the output* of a GenAI model is simultaneously one of the most practically important and most legally unsettled questions in contemporary AI governance. The intuitive assumption (i.e., that the party who prompts a generative model owns the output it produces) is, under current legal frameworks in most jurisdictions, considerably more complicated than it appears. In the United States, this question has now been substantially clarified, though not resolved, by *Thaler v. Perlmutter*. Dr. Stephen Thaler sought copyright registration for an image his AI system, the

"Creativity Machine," had generated autonomously, listing the AI itself as author. The U.S. Copyright Office denied the application; the U.S. District Court for the District of Columbia affirmed that denial in 2023, holding that human authorship is "a bedrock requirement of copyright"; the D.C. Circuit Court of Appeals affirmed unanimously in March 2025; and the Supreme Court denied certiorari in March 2026, exhausting Thaler's appeals. The holding is narrower than it is sometimes described: the courts ruled only that a work generated *entirely* by AI, without human creative contribution, cannot be copyrighted… they did not resolve the far more consequential and fact-specific question of how much human creative contribution is *sufficient* to support a copyright claim in AI-assisted work. The Copyright Office's own guidance draws a rough distinction between AI used as an assistive tool (where sufficient human selection, arrangement, or creative control can support protection) and AI used merely to execute a prompt (which, standing alone, generally cannot) but where any given AI-assisted work product falls along that spectrum remains, in most cases, a matter of legal judgment rather than settled rule.

This situation creates an immediate and practical problem for enterprises using GenAI to produce marketing content, software code, product designs, research reports, and other work product they intend to treat as proprietary. Work product that lacks sufficient human authorship is not eligible for copyright protection and, in the United States, effectively enters the public domain "at the moment" of its creation… meaning that competitors who encounter it may legally copy, adapt, and commercialize it without restriction. Because the threshold of human creative contribution required for different categories of AI-assisted output (currently) remains legally unsettled, business organizations cannot assume that copyright protection attaches automatically to AI-generated work. Building documentation practices that record and demonstrate human creative contribution throughout the AI-assisted production process (e.g., the specific prompts and revisions, the selections made among generated alternatives, or the substantive edits applied) is therefore both a legal risk management practice and, increasingly, a precondition for the organization's ability to assert and defend its intellectual property rights in AI-assisted work product.

The ownership question becomes still more complex when the generative model itself is built on training data whose provenance, licensing status, and copyright clearance are contested. A substantial and growing body of litigation (including suits brought by visual artists against image-generation platforms, by authors against large language model developers, and by news organizations against AI companies that trained models on published journalism) centers on whether training a model on copyrighted material without a license constitutes infringement, and whether the outputs of such models are themselves infringing derivatives of the training data. Courts in the United States and Europe have reached different preliminary conclusions on different aspects of these questions, and the legal landscape will continue to evolve through appellate decisions and, potentially, legislative action over the next several years.

For business organizations, the practical governance implication is immediate regardless of how the litigation ultimately resolves: organizations deploying GenAI systems (whether developed internally, procured from vendors, or accessed through API relationships) should require contractual representations from their AI vendors regarding the provenance and licensing status of training data, the indemnification obligations the vendor assumes in the event of a third-party copyright claim arising from model outputs, and the specific use restrictions that apply to outputs generated through the platform. The absence of these representations in an AI vendor contract is not a minor oversight; it is a material transfer of risk that leaves the enterprise holding legal exposure the vendor (with superior knowledge of its own training data practices) is far better positioned to carry.

Data Leakage into Public Models

The risk of data leakage into public LLMs is perhaps the most immediately actionable, and most frequently underestimated, data governance challenge that enterprises face in their AI adoption journey. Data leakage refers to the inadvertent or uninformed disclosure of confidential, proprietary, or personally identifiable information through its submission as input to AI systems that may use that input to further train or refine their underlying models. This risk is not hypothetical. In 2023, Samsung Electronics disclosed that employees had submitted proprietary semiconductor design specifications, internal meeting notes, and source code to ChatGPT during ordinary productivity use, before the organization had implemented any governance policy for AI tool usage. Under OpenAI's then-applicable terms of service, that submitted information became potential training data for future model iterations… meaning that Samsung's proprietary technical information may have been incorporated into a public model accessible to Samsung's competitors, customers, and any other party with access to the platform. Samsung's experience, now one of the most widely cited enterprise AI governance cautionary tales, almost certainly represents the visible surface of a much larger and largely unmeasured phenomenon: employees across industries routinely submit sensitive client information, internal financial data, personnel records, legal strategy documents, and competitive intelligence to public AI platforms, frequently without awareness that the applicable terms of service may authorize the use of that information for model training.

The governance framework required to manage this risk operates at *three levels* that must function coherently as an integrated system rather than as independent controls:

The *policy* level establishes clear, specific, and enforceable rules defining which categories of information employees may submit to which categories of AI system, under what conditions, and with what documentation requirements. Effective policies at this level distinguish among public AI platforms, enterprise-licensed platforms operating under appropriate data processing agreements, and internally hosted AI infrastructure, and they give employees practical guidance for classifying information and selecting tools accordingly… rather than relying on general principles that leave too much interpretive discretion to individual judgment.

The *contractual and technical* level ensures that enterprise AI platform agreements include data processing addenda that explicitly prohibit the use of submitted data for model training, specify data retention and deletion obligations, establish jurisdictional and technical processing parameters sufficient to satisfy applicable data protection regulation, and provide audit rights adequate to verify compliance. Many enterprise AI providers now offer dedicated agreements with these protections as standard terms, but such agreements are not always the default presented during procurement, and procurement teams without specific AI governance awareness may not recognize their absence as a material gap.

The *technical control* level implements data loss prevention systems, API gateway controls, and AI usage monitoring infrastructure that provide visibility into what information is being submitted to AI systems and enforce policy constraints at the technical layer, rather than relying exclusively on employee compliance with behavioral policy.

Together, these three levels constitute a coherent organizational stance on who controls the organization's data, under what conditions it may be processed by external AI systems, and what governance mechanisms make that control real rather than merely aspirational… a stance best understood as the organization's *data sovereignty posture.*

Data Sovereignty in the Broader Regulatory Sense

Data sovereignty in the regulatory sense adds a further dimension of complexity for organizations operating across jurisdictions. The General Data Protection Regulation (GDPR) in the European Union, the UK GDPR, Brazil's *Lei Geral de Proteção de Dados*, India's Digital Personal Data Protection Act, and a rapidly expanding constellation of national and subnational data protection frameworks all impose requirements on the geographic location of data processing, the legal basis for cross-border data transfers, and the rights of data subjects with respect to automated processing of their personal information... requirements that interact in complex, and sometimes contradictory, ways with the architecture of cloud-based AI platforms that process data across globally distributed infrastructure. An enterprise that routes European customer data through a U.S.-based LLM API without an appropriate transfer mechanism (a Standard Contractual Clause, a Binding Corporate Rule, or an applicable adequacy decision) is not merely taking a legal risk; it is potentially committing a per-violation infringement of data protection law in every jurisdiction whose residents' data is involved, with penalty exposure under GDPR alone of up to four percent of global annual turnover.

The emergence of AI-specific regulatory requirements (including the EU AI Act's obligations around high-risk system transparency, data governance documentation, and human oversight, discussed in the following section) layered on top of existing data protection frameworks creates a compliance architecture of genuine complexity, one that requires legal, technical, and business leadership to collaborate in ways that most organizations' current governance structures are not designed to support. Building that collaborative capacity (through AI-specific legal expertise, cross-functional AI risk committees, and technical infrastructure that makes regulatory compliance the path of least resistance rather than an additional burden) is not merely a cost of doing business in the AI era; rather, it is a condition of doing business responsibly, and increasingly a condition of doing business at all in jurisdictions where enforcement of AI governance requirements is accelerating from aspiration toward reality.

The Global AI Regulatory Architecture: The EU AI Act and Beyond

The regulatory environment governing artificial intelligence underwent a transformation between 2023 and 2026 without precedent in the history of enterprise technology governance. For roughly three decades, the dominant regulatory philosophy governing digital technology in most major economies was permissive by default: new technology categories were allowed to develop, scale, and embed themselves into economic and social infrastructure before regulators intervened, and intervention, when it came, was typically reactive... addressing demonstrated harms rather than anticipated risks. The consequences of that philosophy in data privacy, in social media's effects on democratic discourse, and in algorithmic discrimination in credit and hiring decisions are now well documented, and they have shifted the regulatory posture of governments on both sides of the Atlantic toward a markedly more precautionary approach to AI specifically.

The EU AI Act's Risk-Based Architecture

The European Union's AI Act, which entered into force in August 2024, represents the most comprehensive AI-specific legislative framework yet enacted. Its provisions have applied in stages: prohibitions on unacceptable-risk practices and general AI-literacy obligations took effect in February 2025; governance rules and obligations for providers of *general-purpose AI (GPAI) models* (the large foundation models on which much enterprise generative AI is built) took effect in August 2025, requiring technical documentation, transparency to downstream deployers, and, for models deemed to pose "systemic risk" on the basis of the computational resources used to train them, additional evaluation, incident-reporting, and cybersecurity obligations overseen by the newly established European AI Office. The Act's most operationally demanding tier (obligations for high-risk AI systems under Annex III) was originally scheduled to apply from August 2026. At the time this chapter was written, however, the European Parliament and Council have concluded a legislative "Digital Omnibus" simplification process, formally agreed in stages through May and June 2026, that defers the Annex III high-risk deadline to December 2027 and defers obligations for AI embedded as a safety component in already-regulated products (Annex I — medical devices, machinery, and similar categories) to August 2028. Enterprises should treat this deferral as a matter of prudent planning rather than a reason to delay compliance work: the underlying obligations have not been eliminated, only rescheduled, and the technical documentation, risk management, and human oversight capabilities the Act requires take considerable time to build regardless of the exact deadline in force.

The Act's foundational contribution to global AI governance is its risk-based classification framework, which assigns AI systems to one of four categories (i.e., unacceptable risk, high risk, limited risk, and minimal risk) and calibrates compliance obligations to the severity of potential harm. Understanding this framework well enough to apply it accurately to an organization's specific AI portfolio is the first and most consequential task an enterprise compliance function faces, because misclassification in either direction carries real consequences. Systems incorrectly classified as minimal risk when the Act would designate them high-risk are deployed without the conformity assessments, technical documentation, human oversight mechanisms, and post-market monitoring the Act requires, creating both regulatory exposure and the substantive governance gaps those requirements exist to prevent.

At the prohibited end of the spectrum, the Act bans outright a set of applications whose potential for harm the EU legislature judged too severe for any compliance framework to adequately mitigate: systems that deploy subliminal manipulation techniques to distort behavior in harmful ways; systems that exploit the vulnerabilities of specific groups, including children and persons with disabilities; real-time remote biometric identification in publicly accessible spaces, subject to narrow law enforcement exceptions; AI-based social scoring by public authorities; systems that infer emotional states in workplace and educational contexts; and, added through 2026 amendments, systems used to generate or manipulate non-consensual intimate imagery or child sexual abuse material. Organizations operating in the EU market need to audit their AI portfolios against these categories with genuine rigor rather than as a perfunctory exercise, since several of the prohibited categories are broader in application than their headline descriptions suggest.

The *high-risk category* is where the Act's obligations are most detailed and most consequential for the broadest range of enterprise deployments. The Act designates systems as high-risk either because they serve as safety components of products already covered by EU product safety legislation, or because they operate in one of eight application areas the Act identifies as carrying elevated risk to fundamental rights, health, safety, or access to essential services: biometric identification and categorization; the management of critical infrastructure; educational and vocational systems that determine access to education or evaluate students; employment and worker-management systems, including CV screening, interview assessment, and performance monitoring; systems determining access to essential services, including credit scoring and insurance pricing; law enforcement applications; migration and asylum management; and the administration of justice and democratic processes.

This list will surprise business technology leaders who have not engaged closely with the Act's text, because it captures AI applications already widely deployed (i.e., applicant tracking systems with AI screening components, algorithmic credit decisioning, performance-management platforms, predictive underwriting tools) and subjects them to a compliance regime of considerable rigor and ongoing cost. For systems within the high-risk category, the Act requires a risk management system operating across the system's full architecture and lifecycle; data governance practices ensuring training data is relevant, representative, and free from foreseeable errors; technical documentation sufficient to assess compliance before deployment and to support post-market monitoring; automatic logging enabling post-hoc reconstruction of system operation; transparency to deployers and affected individuals; human oversight measures allowing natural persons to understand, monitor, and override system outputs… commonly distinguished as *human-in-the-loop* (a human reviews individual decisions before they take effect), *human-on-the-loop* (a human monitors ongoing operation and can intervene), and *human-in-command* (a human retains ultimate authority to disable or override the system); as well as demonstrated accuracy, robustness, and cybersecurity appropriate to the system's intended purpose.

Each requirement carries interpretive complexity that legal and technical teams must resolve together, and each carries an implementation cost that should be honestly factored into AI project economics rather than treated as compliance overhead to be minimized after the fact.

Building the Compliance Infrastructure

An enterprise-wide AI *audit and compliance* capability adequate to these obligations (and the parallel obligations imposed by the other regulatory regimes discussed herein) requires organizational architecture that most companies are still in the early stages of constructing. The *foundational* element is an *AI system inventory*: a comprehensive, current register of every AI system the organization develops, deploys, or procures, with sufficient detail about each system's purpose, technical architecture, training data sources, output types, deployment context, and affected populations to support accurate risk classification. This inventory is harder to build than it sounds, because AI capability in large enterprises is frequently distributed across business units, embedded in vendor-supplied software whose AI components are not always transparently disclosed, and continuously evolving in ways that can shift a system's risk classification without triggering a formal review. Keeping the inventory current requires registration of new systems before deployment, change management processes that trigger reclassification review when capabilities or deployment contexts change, and vendor management practices that surface AI components in procured software… itself a substantial program of cross-functional work spanning legal, technology, procurement, and business unit leadership.

The *second* element is *conformity assessment capability*: the internal or externally supported capacity to determine whether specific systems meet applicable regulatory requirements, to document the results in the format regulatory review requires, and to support the third-party audits the Act requires for certain high-risk categories. Third-party conformity assessment for AI is a professional services category still developing its methodologies and market capacity; enterprises that wait until deadlines are imminent to engage this capability will encounter both scarcity and cost premiums that earlier engagement avoids.

Two voluntary but increasingly influential frameworks help organizations operationalize these obligations across jurisdictions rather than building separate compliance architectures from scratch. The *NIST AI Risk Management Framework*, published by the U.S. National Institute of Standards and Technology, organizes AI governance around four functions (i.e., Govern, Map, Measure, and Manage) and provides a widely adopted vocabulary for structuring risk assessment even outside the United States. *ISO/IEC 42001*, the international standard for AI management systems, establishes a certifiable framework analogous to ISO 27001 for information security, giving enterprises a mechanism to demonstrate governance maturity to regulators, customers, and business partners through independent certification rather than self-attestation alone. Neither framework substitutes for jurisdiction-specific legal compliance, but both provide the organizational scaffolding (namely policies, roles, documentation practices, continual improvement cycles) around which jurisdiction-specific obligations can be layered as incremental additions rather than freestanding programs.

Beyond the European Union

The global dimension of AI compliance adds complexity that a purely EU-focused framework cannot address. The United Kingdom, having departed the EU's regulatory framework, has adopted a principles-based, sector-specific approach that assigns oversight to existing sectoral regulators (e.g., the Financial Conduct Authority for AI in financial services, the Information Commissioner's Office for applications involving personal data, the Medicines and Healthcare products Regulatory Agency for AI in medical contexts) and publishes cross-sector AI principles those regulators are expected to operationalize within their existing jurisdiction. This approach is, in some respects, more flexible than the EU's prescriptive requirements, but it demands deeper engagement with sector-specific guidance and closer working relationships with the relevant regulators.

In the United States, federal AI governance remains fragmented between executive branch policy and a growing body of state-level legislation, with a several states (among them Colorado, Illinois, Texas, and California) having enacted or actively developing AI-specific statutory requirements, particularly around automated decision-making in employment and consumer contexts. Because state-level AI legislation is evolving quickly and unevenly, enterprises with U.S. operations should treat this as an area requiring continuous legal monitoring rather than a fixed compliance target. Chinese AI regulation, among the most detailed and rapidly evolving in the world, imposes specific requirements on providers of generative AI services operating in China, including content moderation, watermarking of AI-generated content, real-name user registration, and pre-deployment security assessments… requirements that any enterprise offering AI-powered products in the Chinese market must navigate with counsel specifically experienced in Chinese technology law.

The practical implication for global compliance functions is that a jurisdiction-by-jurisdiction approach (i.e., building entirely separate frameworks for each regulatory regime) is neither efficient nor sustainable as the regulatory landscape continues to expand. The more durable approach identifies the substantive requirements that recur across multiple frameworks (e.g., risk assessment, technical documentation, human oversight, transparency to affected individuals, and post-deployment monitoring) and builds compliance infrastructure around that common foundation, treating jurisdiction-specific obligations as incremental additions rather than as wholly separate programs.

Enterprise AI Governance: Structures, Risk Management, and Accountability

Enterprise-wide AI governance integrates ethics, compliance, and risk management into formal organizational structures. This action requires establishing clear roles and responsibilities, defining policies, and creating oversight mechanisms (e.g., AI ethics committees, review boards, centralized governance functions) that ensure every AI initiative undergoes ethical review, risk assessment, and performance evaluation before deployment and throughout its operational life. Governance of this kind succeeds only when business leaders, technologists, legal counsel, and risk professionals share ownership of it rather than operating in silos; *diffused accountability*, in which responsibility for AI outcomes is spread so broadly that no individual or unit is clearly answerable for them, is itself a recognized governance failure mode, and effective structures deliberately assign responsibility for model development quality, deployment decisions, ongoing monitoring, and the escalation of ethical or regulatory concerns to specific, named roles.

Risk Management and Adversarial Robustness

Risk management in AI extends beyond ethical considerations to encompass operational, reputational, and technical risk. Model risk management ensures that models perform reliably and safely even under changing conditions: organizations must validate inputs, outputs, and underlying assumptions; stress-test models against adverse scenarios; and implement monitoring systems capable of detecting drift or degradation in performance before it produces material harm. Treating deployed models as static assets introduces hidden fragility; treating them as evolving systems subject to continuous validation enables intervention before small errors compound.

Adversarial robustness addresses a related but distinct risk: as AI systems become more influential, they also become more attractive targets for deliberate manipulation. Malicious actors can exploit vulnerabilities in machine learning models by introducing carefully crafted inputs designed to produce incorrect predictions, with consequences ranging from financial loss to operational disruption. Adversarial training, anomaly detection, and systematic robustness testing help defend against such attacks, and resilience against adversarial threats is particularly consequential in high-stakes domains such as autonomous systems, cybersecurity, and financial services, where a single successful manipulation can cause harm disproportionate to the scale of the attack itself.

Documentation, Model Cards, and Algorithmic Impact Assessments

Transparency and accountability are made operational, rather than merely aspirational, through disciplined documentation. Every production AI system should maintain records detailing the business problem it addresses, its model architecture, training data sources and characteristics, preprocessing methods, training procedures, evaluation metrics, known limitations, and ongoing monitoring protocols. This documentation supports auditing and regulatory compliance, but it equally supports internal learning: when decision-making is traceable, organizations can respond quickly to errors, bias, or emerging ethical concerns rather than discovering them only after harm has occurred.

Two documentation artifacts have become standard practice among organizations building mature governance capability. *Model cards* and *system cards* (i.e., structured documentation of an AI system's design decisions, training data characteristics, performance metrics, known limitations, and intended use cases) create a consistent documentation standard across an organization's AI portfolio, supporting internal governance, regulatory compliance, and stakeholder transparency alike. *Algorithmic impact assessments* are systematic evaluations, conducted before deployment, of the potential harms an AI system may cause to affected populations, and they provide a structured methodology for identifying and mitigating risk while it is still comparatively inexpensive to address, rather than after a system is embedded in production workflows and organizational expectations.

Culture, Monitoring, and the Innovation–Compliance Balance

Organizations must adopt continuous monitoring to ensure deployed systems remain ethical, fair, and compliant over time: tracking performance, detecting drift, identifying emerging bias, and assessing ongoing adherence to regulatory requirements. Automated alerting, periodic audits, and human oversight together form a feedback loop that allows organizations to respond proactively to emerging risk rather than reactively to realized harm.

None of this activity is sustainable, however, if it depends entirely on formal review structures. Embedding responsible AI practice requires an organizational culture that treats ethics, risk awareness, and accountability as everyday expectations rather than procedural hurdles. Employees at all levels need enough training to recognize potential bias and enough psychological safety to escalate concerns when they see them; leadership sets these expectations by allocating resources to governance work and by modeling the behavior it expects of others. When responsible AI practice becomes part of an organization's operating culture rather than a compliance overlay applied after the fact, governance shifts from enforcement to enablement… and from a constraint that slows innovation to a discipline that makes innovation durable.

This notion is the heart of the innovation–compliance tension that defines AI leadership today. AI development rewards experimentation and rapid iteration; governance demands discipline and foresight. The organizations that manage this tension most effectively do not treat it as a trade-off to be minimized case by case but instead establish clear risk thresholds and decision rights in advance, allowing teams to innovate freely within well-defined boundaries rather than negotiating governance afresh for every project. The objective is not to slow progress but to ensure that progress compounds rather than destabilizes… that each deployment makes the next one easier to govern well, rather than adding to an unmanaged accumulation of risk.

Data Governance vs. AI Governance

As organizations become increasingly data-driven and AI-dependent, two distinct but deeply related disciplines have emerged as critical pillars of responsible technology management: *Data Governance* and *AI Governance.* The terms are sometimes erroneously used interchangeably in casual conversation, but they address fundamentally different challenges, involve different stakeholders, and guard against different categories of risk. Understanding both (and understanding how they relate to one another) is essential for any organization that collects data, builds models, or deploys AI systems in consequential contexts.

The term data governance refers to the framework an organization uses to manage its data assets with consistency, quality, security, and regulatory compliance. It is, at its core, about ensuring that data (N.B. the raw material of modern business and AI alike) is trustworthy, accessible to those who need it, protected from those who should not have it, and managed responsibly throughout its useful life.

AI governance refers to the framework an organization uses to manage its artificial intelligence systems with accountability, safety, transparency, and regulatory compliance. Governance is, at its core, about ensuring that AI (the decision-making and automation layer built atop an organization's data) behaves as intended, produces outcomes that are fair and explainable, remains under meaningful human oversight, and is deployed responsibly throughout its operational life. Neither discipline is a one-time project; both are ongoing organizational disciplines, and their lifecycles reflect that continuity.

Where the two diverge is in what they ultimately govern, and what risks they are built to catch. Data governance is concerned with the asset itself: is the data accurate, complete, secure, and used within the bounds of consent and regulation? AI governance is concerned with the *behavior built from that asset*: does the model trained on that data make decisions that are safe, unbiased, and aligned with organizational intent… even as the model learns, drifts, or is repurposed in ways no one originally anticipated? A dataset can be perfectly governed (i.e., clean, well-labeled, properly permissioned) and still produce an AI system that discriminates, hallucinates, or behaves unpredictably once deployed. This fact is *precisely* why AI governance cannot simply be an extension of data governance policy; it requires its own controls, including model validation, bias testing, explainability requirements, human-in-the-loop review, and ongoing performance monitoring after deployment.

In practice, the two disciplines are deeply interdependent rather than competing. AI governance relies on data governance as a foundation: an AI system is only as trustworthy as the data on which it was trained and validated, so weak data governance upstream becomes a hidden risk in every model downstream. But good data governance alone offers no guarantee of good AI governance, because the risks introduced at the model layer (e.g., emergent behavior, automated decision-making at scale, or the difficulty of tracing a specific output back to a specific cause) are categorically different from the risks of mismanaged data. Organizations that treat AI governance as a checkbox extension of existing data policy tend to be the ones surprised by biased hiring algorithms, opaque credit decisions, or chatbots making commitments the business never intended to make.

The Data Governance Lifecycle

The process begins with *Data Discovery and Cataloging*… the foundational work of understanding what data an organization in-fact possesses. Before data can be governed, it must be found, inventoried, and classified; organizations frequently discover, at this stage, that their data landscape is far more sprawling and inconsistent than assumed. From there, the process moves to *Data Quality Assessment*, evaluating the accuracy, completeness, consistency, and timeliness of existing data. Poor data quality is one of the most pervasive and costly problems in enterprise technology, and a model trained on low-quality data will produce low-quality outputs regardless of its sophistication… thus making this step foundational to everything that follows.

Policy Definition and Standards come next, establishing the rules by which data will be managed. This phase branches in two directions: *Access Control Rules*, defining who may see and use which data under what conditions, and *Retention Policies*, governing how long different categories of data are kept and when they must be deleted or archived. Together these policies form the regulatory and operational backbone of the governance framework. With policies defined, the organization then moves into *Implementation and Enforcement*... translating rules on paper into controls in practice. Governance frameworks that exist only in documentation provide little real protection; enforcement mechanisms, technical controls, and accountable owners are required to give policy operational meaning.

Compliance Monitoring follows next, as the organization continuously checks whether its data practices conform to both internal standards and external regulation. This phase branches into *Audit Trails*, which create durable records of who accessed or modified data and when, and *Regulatory Checks*, which verify ongoing conformance with applicable law, e.g., GDPR, HIPAA, CCPA and others. The process then advances to *Data Lifecycle Management*, which governs data as it ages: ensuring appropriate archiving, purging obsolete data, and keeping the organization's data estate clean and manageable over time. Finally, the cycle closes (and immediately restarts) with *Continuous Updates*. The regulatory landscape changes, business needs evolve, and new data sources emerge; effective data governance is never finished but *continuously refreshed* to remain relevant and effective.

Data governance exists to protect four core assets: 1) *data integrity* (accuracy and freedom from corruption), 2) *privacy and security* (protection of sensitive information from unauthorized access or exposure), 3) *regulatory compliance* (adherence to applicable law), and 4) *data quality standards* (the ongoing reliability of data as a business asset). Its human ecosystem is correspondingly broad: data stewards own specific datasets and are accountable for their quality and appropriate use; compliance officers ensure the organization meets its legal obligations; IT teams implement the technical controls that make governance real; legal departments interpret regulatory requirements and manage risk exposure; and business unit leaders are simultaneously consumers of governed data and accountable parties for how their teams use it. Organizations implementing data governance *stumble* in predictable ways: namely, treating governance as a one-time project rather than an ongoing discipline, so that policies go stale and controls erode without continuous maintenance; data illiteracy among staff, so that even well-designed frameworks are poorly understood and inconsistently applied; siloed ownership across business units, preventing a coherent governance posture; and weak enforcement, where rules exist on paper but carry no real consequence for violation.

The AI Governance Lifecycle

Where data governance asks, "Is our data trustworthy?", AI governance asks, "Are our AI systems trustworthy?"... a broader, and in some ways more complex, question, because AI systems can cause harm not only through the data they consume but through the decisions they make and the behaviors they exhibit.

The AI governance lifecycle begins with *AI Use Case and Risk Assessment*, evaluating proposed applications before development begins; not all AI use cases carry the same risk profile, and effective governance starts with an honest accounting of what could go wrong and how serious the consequences would be. *Data and Model Inventory* then follows, cataloging the datasets used to train models and the models themselves... just as data governance requires knowing what data an organization has, AI governance requires knowing what models are deployed, what data trained them, and what decisions they influence.

Framework Development next establishes the governance structure within which model development will occur, branching into *Ethical Guidelines*, articulating the values the organization commits to upholding (e.g., fairness, transparency, human oversight, and non-discrimination), and *Bias Detection Protocols*, defining the technical methods used to identify and measure unwanted bias before deployment. With a framework in place, the organization proceeds to *Model Development and Validation*... the construction of AI systems within the guardrails the framework established; governance at this stage is not bureaucratic overhead but the set of practices that makes model behavior predictable and accountable.

Testing and Evaluation then subjects models to rigorous examination before release, branching into *Performance Validation* (does the model achieve its intended objective at acceptable accuracy and reliability?) and *Bias Audits* (are the model's outputs fair across demographic groups and use cases?). A model that performs well on aggregate metrics while systematically disadvantaging a subgroup has failed its governance requirements even when its overall accuracy is high.

Deployment and Monitoring marks the transition from development to production, but governance does not end at launch: models behave differently in the real world than in controlled testing, and ongoing monitoring is essential to catch model drift (the gradual degradation of performance as the real-world distribution of inputs changes) as well as unexpected failure modes that emerge only at scale. The lifecycle closes with *Ongoing Auditing and Optimization*, where models are periodically re-evaluated against both performance and ethical standards, updated as needed, and retired when they can no longer meet governance requirements. Like data governance, AI governance is a *continuous* discipline rather than a deployment checklist.

AI governance protects against four categories of harm: 1) *algorithmic unfairness*, where model outputs systematically disadvantage certain individuals or groups; 2) *opacity*, where neither the organization nor affected individuals can understand why a model reached a particular decision; 3) *unethical use*, where systems are deployed in ways that violate human rights or organizational values; and 4) *biased outcomes*, where historical patterns of discrimination encoded in training data are perpetuated or amplified by model behavior. Its stakeholder community reflects both the technical and social dimensions of this task: AI ethics boards provide organizational oversight and accountability for high-level principles; machine learning engineers and data scientists implement governance practice within the development process itself; compliance teams ensure that deployments meet rapidly evolving regulatory requirements; and critically affected *communities*, the people whose lives are shaped by AI decisions, are increasingly recognized as legitimate stakeholders whose perspectives must inform governance design, not only its aftermath.

Interestingly, AI governance failures follow recognizable patterns. *Afterthought governance* (i.e., bolting ethical review onto systems already built and deployed) is the most consequential, because by that point changing fundamental model behavior is technically difficult and organizationally contentious. *Ignored edge cases* are failure modes that appear rarely in testing but cause serious harm in production. *Poor post-deployment monitoring* means model drift and emergent failures go undetected until damage has already occurred. Lastly, a *lack of diversity* in governance bodies and development teams means blind spots go unchallenged: the people building and reviewing AI systems may systematically fail to anticipate harms that would be obvious to those outside their demographic or professional experience.

Table 10-1: Comparing data governance and AI governance.

Aspect	Data Governance	AI Governance
What it is	Framework for managing data with quality, security, and compliance	Framework for responsible, ethical, and transparent AI deployment
What it protects	Data integrity, privacy, regulatory compliance, quality standards	Algorithmic fairness, transparency, ethical use, protection against biased outcomes
Stakeholders	Data stewards, compliance officers, IT, legal, business leaders	Ethics boards, ML engineers, data scientists, compliance, affected communities

Aspect	Data Governance	AI Governance
Risks addressed	Breaches, poor quality, regulatory violations, access issues, silos	Bias, discrimination, lack of explainability, drift, harm
Common pitfalls	One-time governance, data illiteracy, siloed ownership, weak enforcement	Afterthought governance, ignored edge cases, poor monitoring, lack of diversity

Two Frameworks, One Imperative

In review, data governance and AI governance (Table 10-1) are not competing frameworks; they are complementary ones, and in practice neither is sufficient without the other. An organization with excellent data governance but no AI governance may have clean, well-managed data feeding models that cause discriminatory harm. An organization with thoughtful AI ethics principles but poor data governance will find those principles undermined by the unreliable, biased, or non-compliant data flowing into its systems. Together, these two disciplines constitute the governance infrastructure that responsible AI deployment requires. As AI systems take on increasingly consequential roles across business, government, healthcare, and daily life, the organizations that invest in both (treating governance not as a compliance burden but as a strategic capability) will be the ones best positioned to build AI that is not only powerful, but trustworthy.

Conclusion: Governance as Strategic Capability

Responsible AI, governance, and regulation are not, in the end, three separate obligations layered onto an AI strategy… they are a single discipline viewed from three angles. *Ethics frameworks* articulate what an organization intends its AI systems to do; *governance structures* ensure that intention survives contact with real development teams, real deployment pressure, and real organizational incentives; and *regulation* increasingly requires organizations to demonstrate, with evidence, that the first two are genuinely operative rather than merely declared.

The business case for taking this seriously is both *defensive* and *offensive.* Defensively, inadequate governance creates exposure to regulatory penalty, litigation, reputational damage, and operational disruption that can dwarf the cost of preventive investment… a lesson the EU AI Act's tiered penalty structure, GDPR's "four-percent-of-turnover exposure", and the still-unfolding generative AI copyright litigation all make concrete rather than theoretical. Offensively, organizations that demonstrate genuine commitment to responsible AI (through transparency, fairness, accountability, and security) can differentiate themselves with risk-conscious enterprise customers, with talent, and with regulators, building the organizational trust that enables more ambitious AI deployment over time.

The most sophisticated governance frameworks now being built by leading enterprises treat regulatory requirements not as a *ceiling* to be reached at minimum cost, but as a *floor* from which genuine governance ambition rises. The organizations best positioned as global AI regulation continues to mature (and as enforcement moves from aspiration to routine practice) will be those that have built governance capability deeply enough into their AI development and deployment processes that compliance follows as a natural consequence of good practice, rather than as a separate overlay applied to systems designed without governance in mind. That environment requires a cultural shift already underway in the most capable organizations: from viewing compliance as a *constraint* that slows innovation to viewing it as the *discipline* that makes innovation sustainable; from treating it as a *cost* to be minimized to recognizing it as an *investment* in the organizational trustworthiness that is increasingly a condition of operating at scale in an AI-regulated world.

Undoubtedly, the regulatory environment will continue to evolve. Specific requirements will continue to be refined through enforcement actions and judicial interpretation, and in all probability the compliance burden will increase before it stabilizes. The business enterprises that build genuine governance capability now (e.g., in organizational structure, in technical infrastructure, in legal expertise, and in culture) will navigate that evolution from a position of strength. Those that treat governance as a future problem will find, when the future arrives, that the cost of catching up significantly exceeds the cost of having started earlier.

CHAPTER 11

Leading the AI Digital Transformation

AI Talent, Culture, and Organizational Design for the Agentic Enterprise

ARTIFICIAL INTELLIGENCE IS NOT JUST a technology challenge… it is a human capital and organizational challenge. The success of AI initiatives hinges on the ability of organizations to attract, develop, and retain talent with specialized technical skills, while simultaneously fostering a culture that embraces experimentation, data-driven decision-making, and cross-functional collaboration. Building the right AI *workforce* is foundational to translating technical capability into tangible business value.

Attracting AI talent requires a combination of compelling work, competitive compensation, and a culture that emphasizes learning and innovation. Organizations compete globally for skilled professionals in data science, machine learning engineering, and applied AI research. Tech giants such as Google, Microsoft, and Tesla invest heavily in talent pipelines through partnerships with universities, internal training programs, and global recruiting initiatives. Smaller firms must differentiate themselves by offering challenging projects, clear impact, and opportunities for rapid career growth.

Building an AI-ready workforce involves not only hiring specialists but also developing new capabilities across the enterprise. Data scientists, AI engineers, and product owners remain core roles, but emerging positions are increasingly critical. AI orchestrators, for example, coordinate the work of multiple agent teams, ensuring alignment between autonomous AI systems and business objectives. Agent teams (i.e., groups of AI-driven agents or autonomous systems that collaborate to perform complex tasks) require human supervision and orchestration to function effectively at scale.

Operating models for AI adoption are evolving rapidly. Centralized *Centers of Excellence,* as noted previously, provide standardized tools, governance frameworks, and advanced technical expertise, while federated or decentralized teams embed AI capabilities directly within business units. Hybrid models combine these approaches, allowing core technical teams to set standards while business-aligned teams innovate locally. The key strategy is to balance control and agility, ensuring both enterprise consistency and domain-specific customization.

Cultural adoption is critical for scaling AI. Business organizations must embrace uncertainty and experimentation, recognizing that AI outputs are probabilistic and iterative. Employees need training in AI literacy, including understanding model limitations, data dependencies, and ethical considerations. Leadership plays a crucial role by modeling *data-driven decision-making* and rewarding *evidence-based innovation.* Without a supportive culture, even technically sophisticated AI initiatives may fail to achieve impact.

Organizational change management strategies are also essential to integrate AI into everyday workflows. This element involves clear communication about AI's purpose and benefits, alignment of incentives, and redesign of processes to incorporate AI outputs. Pilot programs and phased rollouts help employees adapt gradually, while feedback loops ensure continuous improvement. Successful change management reduces resistance, accelerates adoption, and maximizes the business value of AI investments.

Measuring AI success requires moving beyond traditional IT metrics to evaluate business outcomes. Key performance indicators (KPIs) should capture both technical performance and results, such as model accuracy and latency, and business impact, including indicators such as cost reduction, revenue growth, customer satisfaction, and operational efficiency. Scorecards provide a holistic view of performance across technical, operational, and strategic dimensions, helping organizations track progress and prioritize initiatives that deliver measurable value.

Business value tracking must account for the full lifecycle of AI initiatives. Models require retraining, monitoring, and adaptation as data, markets, and regulatory conditions change. Continuous measurement ensures that AI systems maintain relevance and effectiveness over time. Integrating KPIs into regular management reporting and discussion fosters accountability and reinforces the alignment of AI initiatives with broader organizational goals.

The rise of autonomous and agentic systems is reshaping the future of work in many ways. AI agents capable of performing complex tasks (from drafting reports to analyzing market trends) augment human capabilities and transform job roles. Employees now increasingly act as supervisors, decision integrators, and exception managers, focusing on judgment, creativity, and strategic thinking, while routine or repetitive work is delegated to AI systems. This shift demands new skill sets, including AI orchestration, data interpretation, and ethical oversight.

Collaboration *between* humans and AI agents introduces new productivity paradigms. Teams now combine human intuition and domain knowledge with AI's computational power, enabling faster decision cycles, more accurate forecasting, and innovative solutions to previously intractable problems. Companies adopting these models gain a *competitive edge* by accelerating insight-to-action timelines and reducing operational inefficiencies across enterprise functions.

AI adoption also prompts organizations to reconsider career paths and workforce development. Continuous learning becomes a central component of employee value, with micro-credentialing, internal academies, and AI-focused mentorship programs emerging as key retention tools. The *workforce of the future* requires adaptability, resilience, and proficiency in leveraging AI as a *partner* rather than a *tool*, reshaping talent management strategies across industries.

Ethical and governance considerations are *intertwined* with workforce transformation. Employees must understand the implications of AI outputs, including fairness, bias, and regulatory compliance. Human oversight remains *critical*, particularly in high-stakes applications such as healthcare, finance, and autonomous systems. Organizations that embed ethics and responsibility into the workflow strengthen trust, reduce risk, and enhance long-term sustainability.

In sum, AI talent, culture, and organizational design are inseparable from technological capability. Building and attracting skilled professionals, creating new roles and operating models, managing change, and implementing robust performance measurement systems are essential for maximizing AI's successful business impact. As AI agents become increasingly autonomous, organizations must cultivate a workforce capable of orchestrating human-machine collaboration, ensuring that AI amplifies human potential and drives sustainable competitive advantage in an emerging agentic world.

Employee Receptivity and the AI Metacognitive Inventory: Diagnosing Organizational Readiness from the Inside Out

Every AI transformation program eventually confronts a truth that no technology roadmap captures and no vendor assessment reveals: the *single most important* variable in determining whether an AI initiative succeeds is not the sophistication of the model, the quality of the data infrastructure, or the size of the investment budget. Rather, it is the *cognitive and psychological readiness* of the human beings who will use, sponsor, champion, and develop the AI systems the business organization deploys.

This readiness (i.e., the combination of mindset, capability awareness, strategic thinking, and ethical judgment that determines how an individual engages with AI as a professional tool and an organizational force) is not uniformly distributed across any workforce. AI *readiness and receptivity* vary dramatically between individuals, in ways that are not reliably predicted by job title, technical background, years of experience, or academic credential. A senior executive with thirty years of industry experience may demonstrate more genuine openness to AI collaboration than a recent technology graduate whose technical confidence has calcified into resistance to AI's encroachment on skills they have only recently mastered. A frontline operations manager without a data science background may possess more sophisticated systems thinking about AI's organizational ripple effects than a data analyst who understands the models but has never been responsible for their organizational consequences.

Understanding this individual variation (and mapping it systematically across the employee population that an AI initiative will touch) represents the diagnostic foundation on which genuinely effective AI organizational change management, champion selection, and organizational development strategy must be built. The *AI Metacognitive Inventory* is the instrument designed for and presented in this book to provide that diagnostic foundation, and its intelligent administration and interpretation is among the most valuable investments that AI transformation leadership can make in the early stages of a major initiative.

The concept of *metacognition* (i.e., thinking about thinking, the capacity to examine and evaluate one's own cognitive processes, assumptions, and limitations) is central to understanding why this inventory is structured as it is and why it produces insights that conventional skills assessments and readiness surveys cannot. A standard AI skills assessment measures *what* an employee knows about AI. A standard change readiness survey measures *how* an employee feels about AI. The *AI Metacognitive Inventory* measures something more fundamental and more predictive of actual AI adoption behavior than either of these: *how an employee thinks about their own relationship to AI…* the accuracy of their self-assessment of their AI capability and limitations, the quality of their judgment about when to trust AI outputs and when to interrogate them, and the depth of their psychological integration of AI as a *legitimate cognitive partner* rather than a threatening interloper or a "magical" solution.

Design and Structure of the Inventory: Four Domains, Fifteen Items

The inventory is organized around four distinct but interconnected domains that together provide a comprehensive map of an individual's AI metacognitive profile. Each domain isolates a different dimension of readiness, and together they reveal patterns that neither technical assessments nor attitudinal surveys can detect on their own. The full instrument is presented in Table 11-1 below.

Instructions: Respond to each statement based on your current initial mindset.

Scale:

1 = Strongly Disagree
2 = Disagree
3 = Neutral
4 = Agree
5 = Strongly Agree

Table 11-1: AI Metacognitive Inventory

Statement		*Rating (1–5)*				
Domain A: Receptivity *Focus: Internal openness to AI as a cognitive partner*						
1.	I view AI as a collaborative "co-pilot" rather than a replacement for human judgment.	1	2	3	4	5
2.	I am comfortable delegating high-stakes analytical tasks to an AI system.	1	2	3	4	5
3.	When AI output contradicts my intuition, my first instinct is to investigate *why* rather than dismiss the result.	1	2	3	4	5
4.	I am willing to adjust my professional identity to include "AI Orchestrator" alongside my core expertise.	1	2	3	4	5
Domain B: Ability *Focus: Understanding the "how" and limits of one's own technical control*						
5.	I can translate a vague business problem into a structured "prompt" or logic-chain that an AI can process.	1	2	3	4	5
6.	I have a consistent, repeatable method for verifying the accuracy of AI-generated insights.	1	2	3	4	5

7.	I can identify the specific point where a business task becomes too complex for current AI capabilities.	1	2	3	4	5
8.	I feel confident in my ability to explain *how* a specific AI tool reached its conclusion to a non-technical stakeholder.	1	2	3	4	5

Domain C: Aptitude
Focus: Seeing the "ripple effects" of AI across an organization

9.	I can visualize how an AI implementation in one department (e.g., Sales) will change data requirements in another (e.g., Supply Chain).	1	2	3	4	5
10.	I can distinguish between an AI "hype" feature and a sustainable competitive advantage.	1	2	3	4	5
11.	I am capable of identifying the "hidden costs" (data cleaning, retraining, API latency) of scaling an AI pilot.	1	2	3	4	5
12.	I prioritize the long-term scalability of an AI solution over immediate, short-term efficiency gains.	1	2	3	4	5

Domain D: Ethics & Governance
Focus: Awareness of risk, bias, and accountability

13.	I proactively scan for data bias or "hallucinations" before presenting AI-driven results.	1	2	3	4	5
14.	I believe the ethical responsibility for an AI's error lies with the business lead, not the software developers.	1	2	3	4	5
15.	I consider the impact of AI on workforce morale and labor structures as a primary strategic variable.	1	2	3	4	5

Total Score (15 items × max 5 = 75 points) ______ **/ 75**

Domain A: *Receptivity* examines the mindset and psychological integration dimension: the individual's internal openness to AI as a cognitive partner rather than a competitor or a tool. The four items in this domain probe trust calibration between AI outputs and human intuition, comfort with delegating analytically consequential tasks to AI systems, the instinctive response when AI output contradicts established expertise, and the willingness to expand professional identity to encompass the role of AI orchestrator alongside domain expertise. Domain A is the foundational layer of the metacognitive profile because the psychological openness it measures is the precondition for everything else… an individual who scores low on receptivity will resist the capability development that Domain B requires, apply the systems thinking of Domain C defensively rather than generatively, and approach the ethical responsibilities of Domain D as compliance requirements rather than genuine leadership obligations.

Domain B: ***Ability*** examines the procedural awareness and verification dimension… not the technical skills the individual possesses, but their metacognitive awareness of their own technical ceiling, their capacity to translate business problems into AI-processable structures, and their possession of a repeatable personal framework for auditing and verifying AI outputs. The distinction between technical skill and metacognitive ability awareness is *critical*: an individual with modest technical skill who possesses accurate awareness of their limitations and a rigorous personal verification framework will make better AI-assisted decisions than one with greater technical skill whose overconfidence in their AI fluency reduces the critical scrutiny they apply to AI outputs.

Domain C: ***Aptitude*** examines the systems thinking and strategic dimension… the individual's capacity to see the organizational ripple effects of AI deployment across departmental boundaries, to distinguish genuine AI competitive advantage from hype-driven feature adoption, to identify the hidden costs of AI scaling before they materialize, and to prioritize long-term AI scalability over short-term efficiency wins. High aptitude scores identify individuals who think about AI not as a tool applied within a function but as a force that reshapes the relationships between functions… and who can therefore bridge the gap between AI program leadership and the organizational business units the program must serve.

Domain D: ***Ethics and Governance*** together examine the responsible leadership dimension… the proactive identification of data bias and model hallucination, the acceptance of business leadership accountability for AI system errors, and the integration of workforce impact considerations into AI strategic planning as a primary variable rather than an afterthought. Domain D scores are among the most diagnostically revealing in champion selection contexts, because the credibility that allows an AI champion to address the legitimate concerns of skeptical colleagues depends significantly on whether those colleagues perceive the champion as someone who takes the ethical and human dimensions of AI seriously or as an uncritical technology advocate.

Scoring, Interpretation, and the Three Readiness Personas

The inventory's fifteen items are scored on a five-point Likert scale (from Strongly Disagree at one to Strongly Agree at five) producing a total score ranging from fifteen to seventy-five and a domain-specific profile that identifies where the individual's metacognitive readiness is strongest and where development investment is most needed. The scoring rubric organizes total scores into three *interpretive personas* that provide immediately actionable characterizations of the individual's overall AI metacognitive profile.

The Skeptical Observer (scores 15–35) describes an individual characterized by high resistance or low AI exposure… someone who tends to view AI as a black box or a professional threat rather than as a manageable organizational asset. This persona does not describe a person without value to an AI initiative; it describes a person whose engagement requires a specific developmental investment: 1) foundational trust building, 2) basic AI literacy development, and 3) carefully sequenced early experiences with AI tools that build confidence and competence simultaneously.

The Emerging Practitioner (scores 36–55) describes an individual with strong receptivity who may lack the systems thinking required to scale AI beyond individual tool usage. This persona can use AI effectively for specific tasks but has not yet developed the organizational and strategic perspective that moves from being a skilled AI user to being an AI workflow designer and organizational change agent. The developmental focus for this persona is the transition from tactical AI application to strategic AI integration… building cross-functional thinking, scalability vision, and governance awareness.

The Strategic Architect (scores 56–75) describes an individual with high metacognitive awareness across all four domains… someone who understands the interplay between technology, organizational design, ethics, and strategy well enough to operate as a genuine AI transformation leader rather than a sophisticated AI user. This is the persona profile most associated with the characteristics of an effective AI champion, and its identification through the inventory is one of the instrument's most strategically valuable applications.

Administering the Inventory: Conditions That Determine Validity

Administering the inventory effectively requires attention to the contextual conditions that determine whether responses generated are genuine metacognitive reflections or strategic self-presentations calibrated to what the respondent believes the organization wants to see. The inventory should be introduced to employees as a developmental tool (i.e., an instrument designed to support their personal AI learning journey and to help the organization provide the most relevant development resources) rather than as a performance evaluation or a selection mechanism.

This framing is not merely motivational; it is analytically essential. An employee who believes the inventory will be used to evaluate their job performance or determine their inclusion in AI initiatives will respond strategically rather than honestly, producing a dataset that reflects organizational impression management rather than genuine metacognitive self-assessment, and that is therefore useless as a diagnostic foundation for organizational change management or champion selection. Individual scores should be visible to the respondent and to AI transformation leadership in a context of explicit developmental support, while aggregate organizational results are reported without individual identification until sufficient trust has been established.

The inventory should be administered at three distinct points in the AI transformation program lifecycle. At *baseline* (before the initiative formally launches) it establishes the metacognitive *starting point* of the employee population and identifies the distribution of personas across the organizational units the initiative will affect. At the *midpoint* (typically six to twelve months after launch) it assesses the metacognitive development that the initiative's early phases have produced and identifies the individuals and units whose development trajectory most warrants additional investment. At the *synthesis point* (meaning the conclusion of the initiative's major phases) it assesses the organizational metacognitive transformation achieved and establishes the new baseline for the program's next development phase.

Champion Selection: The Inventory's Most Consequential Application

The inventory's most strategically consequential application is in the identification and selection of internal *AI champions*… the individuals who will drive adoption, model AI-augmented professional behavior, bridge the gap between the AI program's technical leadership and the business units it serves, and carry the cultural authority within their peer groups to make AI engagement feel like organizational leadership rather than imposed compliance.

The instinct common among technology leaders who have defined the AI program's direction is to select champions from among the individuals who are most technically enthusiastic about AI, who most visibly engage with AI tools, and who most vocally support the program's direction. This instinct is *understandable* and *reliably wrong*. Technical enthusiasm and vocal support are visible and easily rewarded, but they are not the characteristics that make an AI champion effective at driving genuine organizational adoption… particularly among the skeptical, anxious, or ambivalent majority of the workforce whose engagement with AI will determine whether the transformation succeeds or stalls.

The inventory provides an evidence-based alternative to *enthusiasm-based* champion selection by identifying the individuals whose metacognitive profile combines the specific characteristics that genuine championship effectiveness requires: Domain A scores reflecting authentic psychological integration of AI rather than performative enthusiasm; Domain B scores reflecting genuine verification rigor rather than uncritical acceptance; Domain C scores reflecting the organizational systems thinking that allows a champion to connect AI capability to the specific workflow and competitive concerns of the colleagues they will be influencing; and Domain D scores reflecting the ethical and governance awareness that gives a champion the credibility to address the legitimate concerns that skeptical colleagues will raise. This profile (the *Strategic Architect* with strong scores across all four domains) is the metacognitive signature of an effective AI champion.

Domain-Specific Remediation: Designing Targeted Developmental Interventions

Domain-specific analysis of inventory results provides the remediation framework that allows the business organization to design targeted developmental interventions for each of the four metacognitive dimensions identified as requiring investment.

Low Domain A scores (the Receptivity dimension) present what the scoring rubric refers to as a *Human-versus-Machine* mindset, characterized by a view of AI that positions it as a *threat* to professional identity and expertise rather than as a cognitive collaborator. The remediation focus involves augmented intelligence: the reframing of AI's role from replacement to enhancement, communicated not through organizational messaging but through carefully sequenced personal experience with AI tools applied to the specific professional tasks the individual values most and is most expert in. When a Domain A skeptic experiences AI producing a genuinely useful work contribution of which they are proud, the experiential evidence of augmentation is more powerful than any communication campaign.

Low Domain B scores (the Ability dimension) present what the rubric refers to as an *Execution Gap*, or a situation in which receptivity to AI may be strong but the individual lacks the structured verification discipline that prevents over-reliance on AI outputs and the prompt engineering fluency that allows them to structure business problems in AI-processable forms. The remediation focus is the development of a personal AI verification framework… a repeatable, documented process for auditing AI outputs that becomes habitual through practice. This focus is not a technical skill development program; it is a metacognitive discipline development program.

Low Domain C scores (the Aptitude dimension) present what the rubric characterizes as *Siloed Thinking*, or a tendency to evaluate AI applications in terms of their impact within a single function without recognizing the cross-functional data dependencies, organizational ripple effects, and enterprise-wide scalability implications that systems-level AI thinking requires. The remediation focus is exposure to cross-functional AI case analysis… structured experiences that require the individual to trace the organizational consequences of specific AI deployments across departmental boundaries, building the systems intuition that Domain C probes.

Low Domain D scores (the Ethics and Governance dimension) present what the rubric identifies as *Compliance Risk*, or a situation in which the individual's AI engagement is not accompanied by the proactive bias scanning, accountability acceptance, and workforce impact awareness that responsible AI leadership requires. The remediation focus is direct engagement with AI ethics case studies drawn from real organizational contexts… the kinds of documented AI failures, governance gaps, and discriminatory impact incidents that make abstract principles of responsible AI concrete and personally relevant in ways that conceptual ethics training rarely achieves.

Integrating Inventory Insights into Transformation Strategy

The full organizational value of the *AI Metacognitive Inventory* is realized not through any single application of the instrument but through the integration of its insights into the *full architecture* of the AI transformation program's talent strategy. Champion selection grounded in metacognitive profile rather than enthusiasm produces champions who are organizationally credible and developmentally effective. Developmental investment calibrated to specific domain gaps produces learning experiences that are relevant, targeted, and efficient rather than generic and easily dismissed. Communication strategies informed by the distribution of metacognitive personas across the employee population produce messages appropriate to the mindsets manifestly present in the organization rather than those leadership assumes are present.

The longitudinal tracking of metacognitive development across the three administration points provides the organizational learning data that allows the AI transformation program's *human development* component to be continuously refined in response to what the evidence reveals about what is working and what is *not* working. The inventory is, in the most precise sense of the term, a diagnostic instrument… one that reveals the true starting point of the organization's AI transformation journey with a specificity and honesty that no other commonly used assessment approach can match. In a field saturated with technology-centered transformation frameworks that treat human factors as implementation variables to be managed rather than as strategic foundations to be developed, the *AI Metacognitive Inventory* represents a fundamentally different and more realistic approach to the challenge of leading organizations through the most consequential technology transformation of the current era.

The AI Capability Maturity Model: Assessing Organizational Readiness for Enterprise AI Transformation

One of the most persistent and costly mistakes that organizations make at the outset of enterprise AI initiatives is the failure to conduct an honest, structured, and comprehensive assessment of their own *organizational* readiness before committing to the investment, timeline, and capability assumptions that the initiative's business case requires. The consequences of this failure are extensively documented in the AI project failure literature: organizations that launch AI transformation programs without an accurate understanding of their starting point consistently underestimate the foundational work required, overestimate the speed at which AI value can be realized, and discover the gap between assumption and reality at the worst possible moment… when the program is already underway, the investment has been committed, and the organizational expectations have been set at levels that the actual state of the organization's data infrastructure, talent base, cultural readiness, and governance capability cannot support.

The *AI Capability Maturity Model* presented in Table 11-2 is designed to prevent exactly this failure pattern by providing enterprise business leaders with a structured, multi-dimensional diagnostic framework for assessing where their organization is *actually* positioned on the spectrum of AI capability and readiness… not where they *believe* it sits, not where they would *like* it to sit, and not where the most optimistic interpretation of recent AI investments might *suggest* it sits, but where honest, evidence-based assessment against specific and observable organizational characteristics reveals it to be. This assessment is the strategic foundation on which a realistic, appropriately sequenced, and genuinely achievable *AI transformation roadmap* must be built, and its quality (i.e., the honesty, rigor, and analytical completeness with which it is conducted) is among the most reliable leading indicators of whether the AI initiative it precedes will succeed or fail.

Table 11-2: AI Capability Maturity Model

Maturity → Dimension	Initial Level 1 *Initial*	Developing Level 2 *Developing*	Defined Level 3 *Defined*	Managed Level 4 *Managed*	Optimized Level 5 *Optimized*
Strategy	No AI vision or strategy; ad hoc decisions	AI awareness; early discussions but no plan	Defined AI roadmap aligned with business goals	AI initiatives are prioritized and resourced	Continuous AI strategy refinement integrated with business
Data	Data is siloed, inconsistent, and poor quality	Some data consolidation; quality issues remain	Data governance policies established; improving quality	Data is accessible, clean, and managed centrally	Data is continuously improved and leveraged proactively
Technology	Legacy IT systems; limited AI tools	Pilot AI tools tested; infrastructure gaps	AI platforms deployed; IT supports AI projects	Scalable infrastructure; cloud and edge AI adoption	Advanced AI infrastructure; automation and monitoring implemented
Talent	No dedicated AI staff; skills gap in workforce	Some training; limited AI expertise	AI team established; cross-functional training	Skilled AI professionals employed; ongoing training	Talent development embedded; partnerships with academia/industry
Culture	Resistance to change; low AI awareness	Growing interest; some pilot project support	Leadership champions AI; cultural shifts underway	AI-driven decision-making encouraged across teams	AI mindset embedded organization-wide; innovation culture thrives
Governance & Ethics	No formal governance or ethical guidelines	Initial discussions on ethics and compliance	Governance frameworks established; risk assessment	Regular audits; compliance with laws and ethics	Proactive ethical AI leadership; transparency and accountability

The *AI Capability Maturity Model* is constructed on an architectural precedent that has proven remarkably durable across forty years of enterprise capability assessment: the Capability Maturity Model (CMM) and its successor, the Capability Maturity Model Integration (CMMI), developed originally by the Software Engineering Institute at Carnegie Mellon University as frameworks for assessing the maturity of software development processes in defense contracting organizations and subsequently adapted and applied across a wide range of organizational capability domains. The CMM and CMMI frameworks established several design principles that have proven consistently valuable in organizational maturity assessment: the organization of capability into distinct, observable levels that represent qualitatively different states of organizational development rather than merely quantitative differences in resource investment; the assessment of capability across multiple dimensions simultaneously rather than as a single aggregate measure; and the use of maturity level assignments as diagnostic starting points for capability development roadmaps rather than as performance evaluations or competitive rankings.

The *AI Capability Maturity Model* applies these principles to the specific organizational capability domains that enterprise AI readiness requires, producing a framework that is simultaneously rigorous enough to provide genuine diagnostic value and accessible enough to be administered by leadership teams without specialized assessment expertise. Like the CMM before it, the *AI Capability Maturity Model's* value lies not in the precision of the scores it generates but in the structured organizational conversation it initiates… the process of examining specific organizational capabilities against specific observable criteria, identifying the gaps between current state and target state, and developing the sequenced capability development plan that converts assessment insight into transformation action.

The model assesses organizational AI readiness across *six dimensions* that together constitute a comprehensive picture of the organizational capabilities that enterprise AI success requires. The first dimension evaluates *Strategy*… the extent to which the organization has developed a clear, coherent, and business-aligned AI vision, roadmap, and investment framework. An organization at Level 1 on the Strategy dimension has no formal AI vision or strategy, making AI-related decisions on an ad hoc basis in response to individual opportunities or competitive pressures rather than as expressions of a coherent organizational direction. At Level 2, AI awareness has emerged and early discussions are underway, but no formal plan has been developed, and the organization's AI activities remain disconnected from each other and from the business strategy. At Level 3, a defined AI roadmap exists that is explicitly aligned with business goals, providing the strategic framework within which individual AI initiatives are designed and evaluated. At Level 4, AI initiatives are formally prioritized and resourced through the organization's strategic planning and capital allocation processes, and AI investment decisions are made on the basis of portfolio-level strategic logic rather than individual initiative advocacy. At Level 5, AI strategy refinement is a continuous organizational process fully integrated with business strategy development, with the organization's AI capability trajectory actively shaping its business strategy rather than merely responding to it.

The second dimension evaluates *Data*… the quality, accessibility, governance, and strategic management of the organization's data assets. Level 1 organizations have data that is siloed across business units and systems, inconsistent in its definitions and formats, and poor in quality to a degree that makes it unsuitable as a foundation for AI model training or inference without substantial remediation. Level 2 organizations have begun consolidating data across some silos but continue to face significant quality issues that limit its utility for AI workloads. Level 3 organizations have established data governance policies and are making measurable progress on data quality improvement. Level 4 organizations have data that is accessible, clean, and managed centrally through governance frameworks that ensure quality, consistency, and availability for AI workloads. Level 5 organizations treat data as a continuously improving strategic asset, proactively leveraging it to generate insights and competitive advantage rather than managing it reactively as an operational necessity.

The third dimension evaluates *Technology*... the maturity of the organization's IT infrastructure, AI tooling, and operational technology architecture relative to the demands that enterprise AI deployment places on them. At Level 1, the organization's technology environment is dominated by legacy systems with limited AI capability integration and an infrastructure architecture that was not designed to support the data volumes, processing speeds, and API-based integration patterns that AI workloads require. Level 2 organizations have begun experimenting with AI tools through pilot projects but face infrastructure gaps that prevent pilots from being scaled to production. Level 3 organizations have deployed AI platforms that are operational and supported by IT, with an infrastructure capable of supporting the AI projects the organization has prioritized. Level 4 organizations have built scalable AI infrastructure including cloud and edge AI capability, with the architectural maturity to support a growing portfolio of production AI systems. Level 5 organizations operate advanced AI infrastructure with automation and monitoring capabilities that allow AI systems to be managed, updated, and optimized continuously without the manual intervention that less mature infrastructure requires.

The fourth dimension evaluates *Talent*... the availability, development, and retention of the human capability required to design, build, deploy, govern, and continuously improve the organization's AI systems. Level 1 organizations have no dedicated AI staff and face significant skill gaps across the workforce that extend from technical AI development capability to the business-facing AI fluency required for effective AI application and governance. Level 5 organizations have embedded talent development in their operational model, established partnerships with academic institutions and industry ecosystems that provide continuous access to evolving AI capability, and built the internal knowledge management infrastructure that retains organizational AI learning rather than allowing it to walk out the door with individual contributors.

The fifth dimension evaluates *Culture*... the organizational values, leadership behaviors, social norms, and institutional identities that determine how the organization collectively responds to AI-driven change. Level 1 organizations exhibit active resistance to AI-related change and low AI awareness across the workforce — conditions that make effective AI adoption practically impossible regardless of the quality of the technology, data, and talent available to support it. Level 5 organizations have embedded an AI mindset across the entire enterprise, cultivating an innovation culture that actively embraces the continuous change that AI advancement requires and that treats experimentation, learning from failure, and cross-functional collaboration as organizational values rather than aspirational statements.

The (final) sixth dimension evaluates *Governance and Ethics*... the frameworks, processes, accountability structures, and cultural norms that ensure the organization's AI systems are developed and deployed responsibly, in compliance with applicable regulatory requirements, and in accordance with ethical principles that the organization can defend to its employees, customers, regulators, and the broader public. Level 1 organizations have no formal governance framework and no established ethical guidelines for AI — a condition that creates regulatory exposure, reputational risk, and the potential for AI system failures whose consequences extend well beyond the immediate operational context.

Administering the *AI Capability Maturity Model* assessments effectively require a process architecture that balances the analytical rigor required for honest and defensible results with the business organizational accessibility required for broad participation across the leadership and management levels whose perspectives the assessment needs to integrate. The most reliable and most organizationally valuable approach to administration is the structured cross-functional workshop format, in which a facilitated group of senior leaders representing each of the major organizational functions affected by the AI initiative (namely technology, data, operations, finance, human resources, legal and compliance) and the business units that will be primary AI initiative stakeholders together work through the model's six dimensions collaboratively, using the level descriptors to anchor a structured conversation about where the organization currently sits on each dimension. This format is superior to individual survey administration for reasons that are important to understand before designing the assessment process.

First, the most accurate assessment of organizational capability requires the integration of multiple functional perspectives that no single individual, regardless of their seniority or scope of responsibility, can provide from their own experience alone. The Chief Technology/Information Officer has the most detailed view of the Technology dimension but may have a less accurate view of the Culture dimension than the Chief Human Resources Officer, who in turn may have a less accurate view of the Data dimension than the Chief Data Officer, who may have a less accurate view of the Governance dimension than the General Counsel. The cross-functional workshop format brings these complementary perspectives into productive dialogue, producing assessments that are more complete, more accurate, and more organizationally credible than any individual's assessment could be.

Second, the workshop format produces a shared organizational understanding of the assessment results that individual survey administration cannot… because the leaders who participated in the conversation that produced the assessment carry the reasoning and evidence behind each level assignment in a way that makes the assessment actionable in their own functional domains, rather than receiving a report that assigns scores without conveying the organizational understanding that generated them.

The scoring worksheet that accompanies the model provides the structured recording instrument for the assessment, capturing both the numeric level assignment for each dimension on the one-to-five scale and the qualitative justification notes that document the specific evidence, examples, and reasoning that support the assignment. The justification notes are as important as the numeric scores for three distinct reasons.

Initially, they anchor the scores in observable organizational reality rather than abstract aspiration… an organization that assigns itself a Level 3 on the Data dimension because it has established data governance policies must be able to point to specific, observable evidence that those policies exist, are documented, are understood by the people responsible for implementing them, and are actually being applied in the data management practices of the organization rather than residing in a policy document that no operational process references.

Next, they provide the qualitative detail that allows the assessment to be disaggregated across organizational units, business functions, and geographic markets in ways that reveal the internal variation in AI maturity that aggregate organizational scores necessarily conceal… a global organization may have a Level 4 Data capability in its North American headquarters and a Level 2 Data capability in its Southeast Asian operations, and understanding this variation is essential for designing an AI rollout strategy that sequences deployment appropriately rather than treating the organization as uniformly ready or uniformly unready.

Finally, they provide the baseline documentation against which progress can be measured in subsequent assessments, allowing the organization to track not just whether its maturity level scores improve over time but whether the specific capability gaps that the initial assessment identified have been addressed by the development investments the organization has made in response.

Interpreting the assessment results requires analytical discipline that resists two forms of distortion that are predictable features of organizational self-assessment processes. The first form is the *optimism bias* that leads organizations (particularly those in which senior leaders have strong personal investment in the AI initiative's success or have made public commitments to AI transformation timelines) to assign maturity levels that reflect their aspirations for the organization rather than its demonstrated capabilities. The test that guards against optimism bias is specific and demanding: for each level assignment, the assessment team should be able to identify at least three concrete, observable, and independently verifiable examples of the organizational behavior that the level descriptor specifies. If three such examples cannot be identified, the level assignment should be revised downward to the level at which three examples are available.

Another form of distortion is the *anchoring effect* that causes assessment participants to calibrate their level assignments against the organizations they know rather than against the absolute standard the level descriptors define… leading organizations operating in AI-immature industries to assign themselves Level 3 or Level 4 on dimensions where their industry peers are at Level 1 or Level 2, producing scores that are accurate relative to their competitive context but misleading relative to the absolute standard of AI readiness that a successful enterprise AI initiative requires. The remedy for anchoring bias is the explicit and repeated reminder that the capability maturity model's level descriptors represent absolute standards of organizational capability (i.e., standards that are defined by what enterprise AI success requires rather than by what industry peers have achieved) and that a Level 2 score that accurately reflects the organization's starting point is strategically more valuable than a Level 4 score that flatters the organization's current capability while concealing the foundational gaps that will derail the initiative it is designed to support.

Reading the AI Maturity Profile: Strategic Diagnosis, Investment Sequencing, and Longitudinal Tracking

The most strategically important output of the *AI Capability Maturity Model* assessment is not the organization's average score across all six dimensions (a single aggregate number that conceals more than it reveals) but the dimensional profile that maps the organization's specific level assignment on each of the six dimensions simultaneously and reveals the pattern of relative strengths and gaps that should shape the AI initiative's design, sequencing, and investment priorities. Three distinct profile patterns appear with sufficient frequency in enterprise AI assessments to warrant specific strategic guidance for each. The *first* is the *Technology-Forward* profile, in which the organization's Technology dimension score is significantly higher than its Data, Culture, and Governance dimension scores… a pattern that reflects organizations that have invested heavily in AI platform and infrastructure capability without making commensurate investments in the data quality, organizational culture, and governance framework that the technology requires to generate value.

Technology-Forward business organizations are at acute risk of the pilot-to-production failure pattern described in this book… they have the technical capability to build impressive AI demonstrations and the infrastructure to deploy them, but the data quality issues, cultural resistance, and governance gaps that their lower dimension scores reveal will prevent those deployments from generating sustainable business value at scale. The strategic prescription for Technology-Forward organizations is to redirect investment from further technology advancement toward data quality remediation, cultural development, and governance framework construction… building up the dimensions that are constraining the value of the technology investments already made before making additional technology investments that the organizational environment cannot yet absorb.

The *second* profile pattern is the *Strategy-Without-Infrastructure* pattern, in which the Strategy dimension score is significantly higher than the Data, Technology, and Talent dimension scores… an organization that has developed a sophisticated and well-articulated AI vision without building the foundational organizational infrastructure required to execute it. This pattern is common in organizations whose AI strategy has been developed primarily through external consulting engagements that produced high-quality strategy documents without engaging with the organizational realities that determine whether the strategy is executable. The strategic prescription is a phased investment program that systematically builds the infrastructure dimensions to the level that the strategy requires before the strategy's ambitions are communicated to the broader organization… because the gap between an aspirational AI strategy and the organizational capability to execute it, when that gap is visible to the workforce, produces the cynicism about AI transformation that is among the most difficult cultural conditions to reverse.

The *third* profile pattern is the *Balanced-but-Low* profile, in which the organization scores consistently at Level 1 or Level 2 across all six dimensions… a pattern that indicates an organization that is genuinely at the beginning of its AI readiness journey, with no single dimension that provides a strong foundation for rapid advancement. This profile is not a strategic liability; it is an honest starting point, and the strategic prescription it generates is the most straightforward of the three: a sequenced capability development program that begins with the Strategy and Data dimensions (the foundation on which all other dimensions depend) and progresses through Technology, Talent, Culture, and Governance in a sequence that reflects both the logical dependencies between dimensions and the organization's specific competitive and regulatory context.

The *AI Capability Maturity Model* should be administered not as a one-time pre-initiative diagnostic but as a recurring organizational assessment that is conducted at regular intervals throughout the AI transformation program's lifecycle… typically annually for most organizations, and semi-annually for those in the early stages of rapid capability development where the six-month assessment cadence provides sufficiently timely feedback to inform investment and sequencing decisions. The longitudinal tracking of maturity level scores across assessment cycles provides the organizational learning data that allows the AI transformation program to be continuously refined in response to what the evidence reveals about where development investments are generating the most maturity advancement and where the organization's capability development is lagging the initiative's requirements.

It also provides the objective evidence base that AI transformation leaders need to make the case to boards and executive committees for the sustained investment that multi-year AI capability development requires… because a board that can see a coherent picture of maturity advancement across six dimensions over three annual assessment cycles is in a qualitatively better position to evaluate the AI transformation program's progress and make informed investment decisions than one that receives qualitative updates about transformation milestones that provide no structured basis for comparing current capability to the starting point or to the target state that the initiative was designed to achieve. The capability maturity model transforms AI transformation from a narrative about progress into a measurement discipline about capability development… and that transformation, from narrative to measurement, is one of the most reliable markers of the organizational AI maturity level that the model is designed to assess.

CHAPTER 12

The Road Ahead

The AI Horizon: Enterprise Strategy, AGI Implications, and the Road Ahead

AS THE AI REVOLUTION ACCELERATES, the coming decade promises to bring unprecedented shifts in enterprise capabilities, business models, and workforce dynamics. While today's AI systems excel in narrow tasks (e.g., predicting maintenance needs, personalizing marketing, or optimizing logistics) the horizon toward *Artificial General Intelligence (AGI)* introduces both transformative opportunities and complex uncertainties. Organizations must prepare not only to adopt AI technologies at scale but also to adapt their strategy, culture, and governance frameworks to the emerging realities of the 2026--2036 period.

The pursuit of AGI (systems capable of reasoning, learning, and generalizing across tasks at or above human-level proficiency) has long been considered the ultimate frontier of artificial intelligence research. While AGI remains largely speculative, multiple pathways are being explored by industry and academia. One approach emphasizes scaling existing deep learning architectures to extreme sizes, leveraging massive datasets and compute resources to gradually close the gap between narrow AI and general reasoning. Another approach focuses on *hybrid models* that integrate symbolic reasoning, causal inference, and structured knowledge graphs with probabilistic learning, enabling AI to perform abstract reasoning tasks more reliably. A third approach emphasizes multi-agent systems, where collections of specialized AI agents collaborate to solve complex, open-ended problems. Each approach carries distinct implications for business, from automating complex knowledge work to creating entirely new markets that depend on AGI-level capabilities.

The business implications of AGI are profound. Firms that develop or gain early access to AGI-like systems could redefine competitive boundaries by performing strategic tasks that today require human experts—advanced research and development, global supply chain orchestration, and high-level financial strategy, among others. These capabilities could compress innovation cycles, create hyper-efficient operational models, and unlock new avenues of value creation. However, AGI also poses systemic risks: unintended behaviors, ethical dilemmas, regulatory scrutiny, and profound labor displacement. Organizations must therefore integrate robust governance frameworks and scenario planning to mitigate potential negative outcomes while harnessing the advantages of advanced AI.

Even short of full AGI, the next five years will see *rapid maturation of narrow AI across enterprises*, with increasingly sophisticated, context-aware models deployed across functional areas. Adoption scenarios range from conservative incrementalism (where AI augments existing processes with modest efficiency gains) to aggressive transformation, in which entire business models are reimagined around predictive, autonomous, and agentic systems. A mid-range scenario envisions a hybrid approach, with AI embedded deeply in certain functions while others remain human-led, gradually building confidence and organizational competency before full-scale adoption. Understanding these scenarios helps executives prioritize investment, manage risk, and design flexible AI strategies that can pivot as technology and markets evolve.

Companies that successfully navigate AI adoption over the next five years will share several characteristics. First, they demonstrate *AI fluency at the leadership level*, enabling decision-makers to interpret model outputs, understand probabilistic reasoning, and balance technical possibilities with business realities. Second, they maintain *robust data infrastructure*, ensuring high-quality, accessible, and ethically sourced data for all AI initiatives. Third, they adopt *iterative, experimental deployment processes*, embedding continuous monitoring, model retraining, and feedback loops as standard practice. These principles form the foundation of what can be termed the AI-ready organization.

A key component of AI readiness is *organizational design*. Enterprises will increasingly adopt federated or hybrid AI structures, combining centralized Centers of Excellence for advanced research, governance, and infrastructure with embedded teams that develop domain-specific models and solutions. This configuration allows firms to maintain consistency, promote innovation, and accelerate deployment across multiple business units. Alongside structure, a *culture that embraces experimentation and uncertainty* will differentiate successful adopters. Probabilistic models require tolerance for failure and iterative improvement; organizations that reward learning from experiments rather than punishing risk-taking will scale AI more effectively.

The next decade will also highlight the importance of *responsible AI governance.* Ethical considerations, fairness, and explainability will remain central, especially as AI systems increasingly interact with human lives and decisions at scale. Governance boards, model audits, and ethical review processes must evolve to account for new types of AI risk, including algorithmic drift, bias amplification, and adversarial vulnerabilities. Enterprises that invest early in these capabilities will not only comply with emerging global regulations (e.g., the EU AI Act revisions or new U.S. federal AI oversight) but also build trust with customers, partners, and employees.

From a technological perspective, *integration of AI into business operations will continue to expand.* Predictive maintenance, personalization, fraud detection, and supply chain optimization will mature, while natural language and reasoning models will increasingly support complex decision-making tasks, such as contract analysis, R&D hypothesis generation, and strategic forecasting. Companies will learn to combine human judgment with AI recommendations, leveraging the complementary strengths of humans and machines in hybrid workflows. In many cases, these applications will move from pilot projects to core operational capabilities, creating measurable improvements in efficiency, revenue, and risk mitigation.

Scalability will remain a central challenge. High-performing AI models demand compute, storage, and energy resources that must be managed cost-effectively. Organizations that invest in cloud-native architectures, modular pipelines, and MLOps best practices will have the agility to deploy AI rapidly while controlling operational expenses. AI will increasingly shift from a *capital-intensive (capex) to an operationally managed (opex) cost structure*, aligning financial models with usage-based or token-based pricing for inference, which can accelerate adoption and ROI tracking.

Another emerging theme will be *AI-enabled strategic differentiation.* Companies with proprietary datasets, advanced models, and integrated ecosystems will establish defensible AI moats, limiting the ability of competitors to replicate their advantages. Strategic partnerships, joint ventures, and selective acquisitions may become critical mechanisms to accelerate AI adoption or gain access to frontier models without full in-house development. Business organizations that fail to act proactively may face competitive disruption, as seen in the first-wave adoption of recommendation engines, autonomous logistics, and fraud detection.

The Gartner AI Hype Cycle: Implications for the Future of AI in Business

Few analytical frameworks have proven more consistently useful (and more consistently misapplied) in the history of enterprise technology management than the Gartner Hype Cycle®, a model used to describe the maturity, adoption, and social application of emerging technologies. Since its introduction in the mid-1990s, the framework has provided technology leaders and business strategists with a deceptively simple but genuinely powerful model for understanding the relationship between technological expectation and technological reality across the lifecycle of an emerging innovation. The model's five phases (i.e., the *Technology Trigger* that initiates public awareness, the *Peak of Inflated Expectations* that follows early enthusiasm, the *Trough of Disillusionment* that arrives when reality fails to match expectation, the *Slope of Enlightenment* along which practical understanding and genuine capability accumulate, and the *Plateau of Productivity* where mainstream adoption delivers stable and predictable value) have described the trajectory of enough significant technologies, from the early internet to blockchain to virtual reality, that dismissing the framework as oversimplified would be a mistake.

The more important question for business leaders navigating the current AI landscape is not whether the Hype Cycle framework is valid in the abstract. The question is where artificial intelligence (or more precisely, the multiple distinct AI technology categories that are simultaneously progressing through their own independent hype cycle trajectories) currently sits within the framework, what the historical pattern of prior technology hype cycles predicts about the path ahead, and what strategic and organizational implications those predictions carry for enterprises making consequential AI investment decisions today. The answers to these questions are neither uniformly optimistic nor uniformly cautionary. They are, in the way of genuinely useful strategic analysis, considerably more nuanced and considerably more actionable than the binary enthusiasm and skepticism that currently dominate most AI business discourse.

The first and most important analytical complication in applying the Hype Cycle framework to contemporary AI is the recognition that artificial intelligence is not a single technology progressing through a single hype cycle, but a *portfolio* of distinct and unevenly mature technologies each of which is at a different point in its own cycle simultaneously. Large language models and generative AI, which captured the overwhelming majority of enterprise and public AI attention between 2022 and 2025, moved through the Technology Trigger and the Peak of Inflated Expectations with a speed and intensity that had few precedents in the history of enterprise technology… the period between ChatGPT's public launch in November 2022 and the peak of generative AI investment and media attention in mid-2023 represented perhaps the fastest ascent to a hype cycle peak ever recorded for an enterprise technology category. By 2024 and into 2026, the characteristic dynamics of the Trough of Disillusionment were clearly visible in the enterprise AI landscape: a growing body of documented evidence that generative AI pilot results frequently failed to translate into enterprise-scale value, increasing scrutiny of AI investment returns from boards and CFOs who had approved substantial AI budgets on the basis of pilot enthusiasm that operational deployment had not validated, a series of high-profile AI deployment failures and reputational incidents that introduced caution into boardroom AI conversations that had previously been characterized by uncritical enthusiasm, and a measurable moderation in the most extravagant predictions about AI's near-term economic impact that had characterized the peak period. This does not represent a failure of AI as a technology category. It represents the entirely predictable and historically consistent process by which a powerful technology category transitions from the distorted expectations of the hype peak toward the more accurate and more useful understanding of genuine capability and genuine limitation that precedes mainstream productive adoption. The Trough, for all its negative connotations, is where the most important organizational learning about a technology happens — and the enterprises that use it productively, investing in capability building and governance infrastructure while their less disciplined competitors retreat from AI investment in response to disappointing pilot returns, are precisely the ones that emerge strongest when the Plateau of Productivity arrives.

The historical precedents that the Hype Cycle framework offers for understanding AI's trajectory are both instructive and sobering in ways that the current AI discourse rarely acknowledges with sufficient seriousness. The internet provides perhaps the most directly relevant comparison. The commercial internet's hype cycle peaked in the late 1990s with valuations, investment levels, and productivity predictions that, viewed from the present, appear almost delusional in their disconnection from what the technology could deliver at that stage of its development. The subsequent collapse of the dot-com bubble sent the technology through a Trough of Disillusionment of dramatic depth and duration… destroying trillions of dollars of market value, ending thousands of companies, and producing a period of deep institutional skepticism about internet-based business models that persisted well into the mid-2000s. And yet the internet's actual economic impact (on commerce, on communication, on media, on finance, on every dimension of business and social organization) ultimately exceeded even the most extravagant predictions of the hype peak, not in the timeframe that peak enthusiasm anticipated but in a longer arc of transformation that unfolded over two decades rather than two years. The lesson that this history offers for AI is not that the current enthusiasm is wrong about AI's ultimate significance. Rather, it is that the timeframe assumptions embedded in peak-period predictions are almost always wrong… compressed by enthusiasm into a near-term horizon that systematically underestimates the organizational, infrastructural, and human capability development that translates technological potential into economic reality. The enterprises that built genuine internet capability during the trough period (e.g., investing in e-commerce infrastructure, digital customer experience, and data architecture while the broader market was retreating from internet investment) were the ones that captured disproportionate value when the Plateau of Productivity arrived. The parallel implication for AI is direct and strategically significant.

Agentic AI systems (i.e., autonomous AI agents capable of planning, tool use, multi-step reasoning, and independent action across complex workflows) represent the technology category whose hype cycle position is perhaps *most consequential* for enterprise AI strategy over the next three to five years. As of 2025, agentic AI sits near the Peak of Inflated Expectations on its own hype cycle trajectory, with enterprise enthusiasm running well ahead of the governance frameworks, reliability standards, and organizational capabilities required to deploy autonomous AI agents responsibly at scale. The failure modes that agentic AI systems exhibit (i.e., cascading errors in multi-step reasoning chains, unpredictable behavior in novel situations outside the training distribution, security vulnerabilities to prompt injection and adversarial manipulation, and the fundamental challenge of maintaining meaningful human oversight over systems designed to operate with increasing autonomy) are not yet fully understood, and the organizational and regulatory infrastructure required to manage them does not yet exist in mature form. This notion does not mean that agentic AI will not deliver the transformative enterprise value that its proponents describe. The underlying trajectory of capability development strongly suggests that it will, over a timeframe measured in years rather than decades. It means that enterprises deploying agentic AI systems at scale before that infrastructure exists are operating in the zone of maximum hype cycle *risk*… where the gap between anticipated and actual value is widest, where the failure incidents that accelerate the Trough are most likely to occur, and where the reputational, regulatory, and financial costs of those failures fall disproportionately on early movers who deployed without adequate governance. The strategically optimal position for most enterprises (with the exception of those whose competitive situation makes first-mover risk genuinely necessary) is to build agentic AI capability now, deploy it in controlled and well-governed contexts, and scale aggressively as the reliability, governance, and organizational readiness conditions that sustainable deployment requires are progressively established.

The Plateau of Productivity that the Hype Cycle framework predicts as AI's eventual destination carries implications for business organization, competitive dynamics, and economic structure that are worth examining carefully, because the Plateau of AI productivity will look fundamentally different from the plateaus that prior enterprise technology categories reached. When cloud computing reached its Plateau of Productivity, it delivered substantial and durable efficiency gains to the enterprises that adopted it well, but it did not structurally transform the competitive dynamics of most industries… it lowered infrastructure costs broadly, creating efficiency gains that were relatively widely distributed across adopters rather than concentrating advantage in a small number of winners. The Plateau of AI productivity is *unlikely* to follow this pattern, for reasons rooted in the specific economics of AI advantage. AI systems improve with data, and data accumulates with use, creating the *compounding* advantage dynamics that this book has described as AI moats… structural competitive positions that widen over time rather than converging toward industry parity as conventional technology adoption curves predict. The enterprises that arrive at the AI Plateau of Productivity with the largest proprietary datasets, the most refined models, the deepest organizational AI fluency, and the most sophisticated human-AI collaboration practices will not simply be more efficient than their less advanced competitors. They will be operating with a fundamentally different production function (one that allows them to deliver better products, make better decisions, serve customers more effectively, and respond to market changes more rapidly at lower incremental cost) in ways that compound rather than converge over time. This difference means that the strategic cost of arriving late to the AI Plateau is not merely a temporary efficiency disadvantage of the kind that technology adoption lags have historically produced. It may be a structural *competitive* disadvantage of the kind that is very difficult and very expensive to close once it has opened to sufficient width.

The most practically useful implication of the Hype Cycle framework for enterprise AI strategy is perhaps the simplest and the most consistently ignored: the Trough of Disillusionment is not a signal to reduce AI investment. Instead, it is the optimal moment to increase it selectively and strategically. The enterprises that will define the AI Plateau of Productivity (meaning those that will have built the data assets, organizational capabilities, governance infrastructure, and human-AI collaboration practices that constitute genuine and durable AI advantage) are not the ones that rode the hype peak with the largest announcement budgets and the most ambitious transformation proclamations. Rather, they are the ones that maintained disciplined, sustained investment through the Trough, using the period of reduced competitive pressure and more realistic technology pricing to build the foundation that hype-peak conditions made too expensive and too chaotic to construct properly.

The historical record of technology hype cycles is remarkably consistent on this point: the Trough *rewards* patience, discipline, and genuine capability building, and *punishes* both the premature enthusiasm of the peak and the premature retreat that the Trough itself tempts. For the business leaders, technology executives, and organizational strategists reading this book, the Gartner AI Hype Cycle offers its most valuable guidance not as a prediction of where AI will go (i.e., the ultimate destination of mainstream productive adoption is not seriously in doubt) but as a map of where the journey will pass through on the way there, and a reminder that the organizations best positioned to thrive at the destination are those that navigate the journey with the clearest understanding of where they actually are on the map at any given moment.

The Productivity J-Curve: Managing the Value Dip During AI Implementation

As noted previously, Erik Brynjolfsson's contribution to the economics of technology adoption is, in the context of the current AI transformation wave, less a historical observation than a living prediction… one that is being validated in real time across thousands of enterprise AI programs whose financial results are diverging from their business case projections in patterns that are entirely consistent with what Brynjolfsson and his colleagues identified in their study of information technology investments in the 1980s and 1990s. The *productivity paradox* that Brynjolfsson famously articulated (i.e., the observation that the period of most intense investment in information technology was associated not with accelerating productivity growth but with a sustained productivity slowdown that persisted for more than a decade before the expected gains materialized) was initially received as a puzzle, a seeming contradiction between the evident power of computing technology and the stubborn absence of that power in aggregate productivity statistics.

The resolution of the paradox, which Brynjolfsson and his collaborators developed over subsequent years of research, is now understood as one of the most *important* insights in the economics of technological change: the productivity gains from general purpose technologies are not realized at the moment of technology adoption but at the moment of organizational transformation, and the gap between those two moments (during which investment costs are fully realized but productivity gains have not yet materialized) is the J-curve, the value dip, the productivity "trough" that every organization deploying transformative technology must pass through before it reaches the upward slope of genuine return. For business leaders navigating enterprise AI transformation today, this insight is not an academic abstraction. It is a practical prediction about the financial and organizational trajectory their AI programs will follow, a warning about the specific moment at which that trajectory will be most vulnerable to the short-term pressures that cause organizations to abandon transformative investments before they reach the inflection point where returns begin to compound, and a roadmap for the organizational and management practices that determine whether an enterprise passes through the value dip successfully or gets trapped in it *indefinitely*.

The J-curve concept derives its name from the shape of the return profile that transformative technology investments characteristically produce when plotted against time. The curve begins above the baseline… at the moment of investment decision, the organization's financial position reflects the cost of the AI program without yet reflecting any of its benefits, producing an immediate negative return that represents the initial downward stroke of the "J." As implementation proceeds, this negative return deepens rather than improving, because the full costs of transformation (e.g., technology infrastructure, talent acquisition and development, process redesign, change management, organizational restructuring, and the productivity disruption that accompanies any significant change to established working practices) are realized progressively throughout the implementation period, while the benefits remain largely unrealized because they depend on the organizational capabilities, redesigned processes, and human-AI collaboration practices that are still being built. This is the bottom of the J (i.e., the value dip) and it is the moment of maximum financial and organizational stress, maximum leadership doubt, maximum board and investor pressure to reduce investment, and maximum probability of the premature abandonment that converts a temporary performance trough into a permanent failure.

Organizations that survive the bottom of the J and continue investing through it eventually reach the inflection point… the moment at which the organizational transformation required to realize AI's productivity potential has proceeded sufficiently that returns begin to materialize and compound, producing the upward stroke of the J that ultimately delivers returns that exceed not just the cost of the investment but the returns that would have been available from any alternative deployment of the same capital. The shape of the curve (namely the depth of the dip, the duration of the trough, and the steepness of the eventual recovery) varies significantly across organizations and AI program types, but the fundamental pattern is remarkably consistent, and understanding it in sufficient depth to manage through it rather than be managed by it is among the most important capabilities that AI transformation leadership requires.

The specific mechanisms that produce the value dip in AI transformation programs are worth examining in detail, because understanding them is a prerequisite for managing them. The *first* and most immediate mechanism is the direct cost of AI infrastructure investment… the capital and operational expenditure on cloud computing, data platforms, AI development tooling, and the talent required to build and maintain AI systems, all of which are incurred upfront or early in the transformation program before any of the AI systems built on that infrastructure are generating business value. This cost is visible, measurable, and appears directly in financial statements in ways that create immediate pressure on the program's apparent ROI.

The *second* mechanism is the productivity disruption cost, meaning the well-documented decline in individual and team productivity that occurs during the period when employees are transitioning from established working practices to new AI-augmented workflows. This disruption is not a sign that something has gone wrong. It is an inevitable consequence of the learning investment that genuine capability development requires, and it is entirely analogous to the productivity dip that accompanies any significant change in tools, processes, or organizational structure. A legal team transitioning from manual contract review to AI-assisted review will, during the transition period, be less productive than it was before... because its members are simultaneously doing their jobs and learning a new way of doing their jobs, and the cognitive overhead of that simultaneous learning and working reduces output quality and quantity in ways that are temporary but real.

The *third* mechanism, and the one most specific to AI transformation as distinct from prior technology adoption cycles, is the reorganization cost... the investment required to redesign the business processes, organizational structures, performance frameworks, and management practices that must change for AI capability to generate its full potential value rather than being layered on top of unchanged organizational structures that prevent it from doing so. Brynjolfsson's research established conclusively that the productivity gains from information technology were not generated by technology adoption per se but by the organizational complementary investments that effective technology adoption requires... and that organizations that made the technology investment without the organizational investment captured only a fraction of the available returns. The same relationship applies with even greater force to AI, where the gap between the potential value of AI capability and the organizational capability required to realize it is wider and more consequential than it was for any prior enterprise technology category.

The duration and depth of the value dip varies across organizations in ways that are not random but are systematically related to identifiable organizational characteristics and program management decisions that business leaders can influence. The depth of the dip is primarily determined by the ratio of organizational investment to technology investment in the AI program... organizations that invest heavily in technology while underinvesting in the process redesign, change management, training, and organizational restructuring required to make technology adoption genuinely productive experience deeper and longer dips than those that balance these investment categories appropriately. The duration of the dip is primarily determined by the speed at which the organization can build the complementary human and organizational capabilities that AI value realization requires... which in turn depends on the quality of change management, the depth of leadership commitment, the effectiveness of talent development programs, and the organizational culture's capacity for the kind of continuous learning and adaptation that AI transformation demands.

Research on prior technology transformation programs (e.g., the IT productivity paradox, the adoption of lean manufacturing, the transition to cloud infrastructure) consistently finds that the organizations that emerge from the value dip most quickly and with the strongest return trajectories are not those that spent the most on technology or moved the fastest on deployment, but those that invested most deliberately in building the organizational capabilities that the technology required and that maintained their investment commitment most consistently through the period of maximum financial and organizational stress. The J-curve, in other words, does *not* reward the fastest movers or the largest spenders. It rewards the most organizationally disciplined… the enterprises that understood what the curve predicted before they entered it, prepared their boards and stakeholders for the dip before it arrived, and maintained the investment and organizational commitment required to reach the inflection point rather than retreating under pressure at precisely the moment when persistence would have been most valuable.

The management implications of the J-curve framework for enterprise AI transformation are specific, practical, and frequently the opposite of the instincts that short-term financial pressure tends to produce. The *first* and most important implication is the necessity of honest expectation setting with boards, investors, and senior leadership before the AI transformation program begins… communicating explicitly that the financial trajectory of the program will include a period of negative or below-target returns that is not a sign of program failure but a *predicted and necessary* phase of genuine transformation, and establishing the governance framework that will distinguish a healthy value dip from a genuine program failure requiring intervention. Organizations that make this communication investment before the dip arrives are vastly better positioned to maintain the investment commitment and organizational morale required to pass through it than those that present boards and stakeholders with AI business cases that project a smooth upward trajectory from investment to return… because when the dip arrives, as it inevitably will, the absence of prior communication about it is interpreted as evidence that something has gone wrong, triggering exactly the pressure to reduce investment and slow transformation that turns a manageable trough into a project-ending crisis.

The *second* implication is the critical importance of early wins… deliberately designed AI deployments that generate visible, measurable value quickly enough to demonstrate genuine progress during the trough period, maintaining organizational confidence and stakeholder support while the deeper organizational transformation required for full value realization proceeds. Early wins are not substitutes for genuine transformation; they are the organizational fuel that sustains the commitment to transformation through the period when the financial returns from deeper change have not yet materialized. Selecting, designing, and communicating early wins with strategic intentionality (i.e., choosing use cases that are genuinely valuable, genuinely achievable in the near term, and genuinely representative of the broader AI capability the organization is building) is a core leadership competency for AI transformation programs that most organizations approach too casually.

The *third* implication concerns the measurement framework that governs the program throughout the value dip period: organizations that measure AI program success exclusively through financial return metrics during the trough will consistently misread the evidence, interpreting the healthy organizational investment that is building the foundation for future returns as unproductive cost rather than value-generating capital formation. Complementing financial return metrics with leading indicators of organizational capability development (e.g., AI literacy measures, process redesign completion rates, model performance and adoption metrics, and the human-AI collaboration quality indicators described in the intangible capital section of this book) provides the measurement architecture that allows leadership to distinguish an organization that is progressing through the J-curve successfully from one that is genuinely failing to build the organizational foundation that the curve's upward stroke requires.

The J-curve framework carries one *final* implication that is perhaps the most important for the long-term competitive positioning of enterprises navigating AI transformation, and it is one that connects the productivity paradox directly to the competitive dynamics that this book has examined throughout. The returns that materialize on the upward stroke of the AI productivity J-curve are not simply the recovery of the value invested during the dip. They are, for organizations that have made the organizational transformation genuinely and completely, compounding returns of the kind that Brynjolfsson's research identified as the characteristic signature of successful general purpose technology adoption… returns that grow over time as the organizational capabilities built during the transformation generate increasing value through their interaction with continuously improving AI technology, accumulating proprietary data assets, and deepening human-AI collaboration practices.

The organizations currently navigating the value dip of their AI transformation programs are not simply enduring a temporary financial inconvenience on the way to recovering their investment. They are, if they are managing the transition with the organizational discipline and investment commitment that the J-curve framework prescribes, building the foundation of a compounding competitive advantage whose full magnitude will only become visible from the upward stroke of the curve… a vantage point that will, by the time it is reached, be available only to those who committed to the journey before the destination was certain and maintained that commitment through the moments when uncertainty was highest and pressure to retreat was strongest. That is, ultimately, what strategic leadership in the AI era requires: not the ability to predict which specific technologies will prevail or which use cases will generate the greatest returns, but the organizational wisdom to understand the *shape* of the transformation journey, the courage to begin it honestly, and the discipline to complete it fully.

Two Curves, One Moment: Reading Sentiment and Productivity Together

The two frameworks examined in the preceding sections describe the same turbulent stretch of enterprise AI adoption, and when their trajectories are plotted against the same time axis they produce a visual that invites a natural but misleading inference: that the Gartner Hype Cycle and Brynjolfsson's productivity J-curve are simply two versions of the same story, one told by analysts and one told by economists. As noted in Figure 12.1 both curves dip. Both curves eventually recover. Both have been invoked, often interchangeably, to warn executives against the twin errors of premature abandonment and premature triumphalism. But the resemblance in shape conceals a difference in mechanism that matters *enormously* for how a business leader should act on either curve, and that difference (not the surface similarity) is the more useful story for the reader of this book to carry forward.

Rogers' diffusion-of-innovation framework, which underlies the Gartner Hype Cycle, is a curve of adopter psychology and market narrative. It tracks how a population of potential adopters (i.e., innovators, early adopters, an early majority, a late majority, and laggards, in Rogers' original taxonomy) forms and revises its collective beliefs about a technology's value, fueled by media coverage, vendor promises, analyst commentary, funding announcements, and the visible successes and failures of early movers. The Hype Cycle is, in this sense, a *sentiment index*: it measures what a market believes and says about a technology, not what that technology has yet produced in a company's financial statements. Brynjolfsson's J-curve is a different kind of instrument entirely. It is an *economic* measurement curve, built from productivity statistics, firm-level financial performance, and controlled comparisons of matched investment, and it tracks a specific accounting reality… the gap between capital deployed on a general-purpose technology and the measured output that technology has, at any given moment, in-fact generated. One curve is a record of *collective perception*; the other is a record of *realized value*. They can move together, they can move apart, and mistaking one for the other is precisely the error this section is meant to correct.

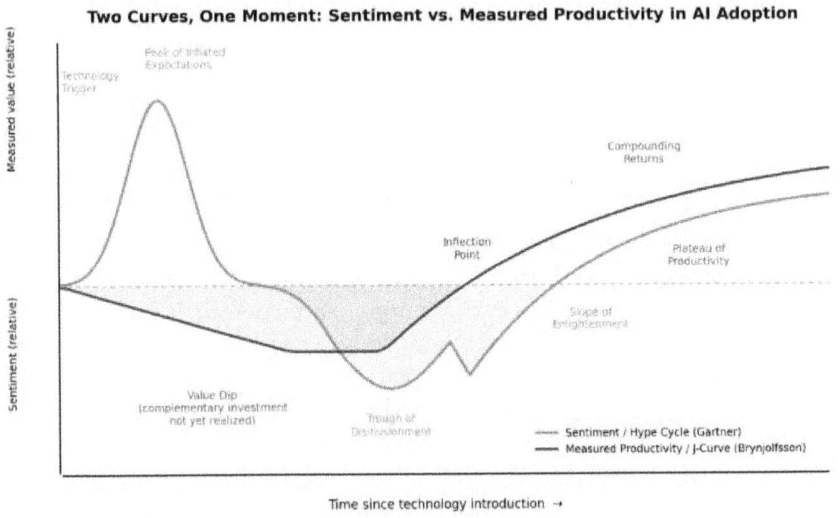

Figure 12-1: The Hype Cycle tracks sentiment; the J-curve tracks measured productivity. Both dip and recover on a similar timeline, but for different reasons.

The reason the two curves nonetheless trace such similar silhouettes over similar timeframes is not coincidence, and the clearest way to see why is through a piece of economic history that predates AI, predates the internet, and even predates the computer: the electrification of American manufacturing. Economic historian Paul David's influential analysis of the factory electric dynamo showed that manufacturers adopted electric power in the 1880s and 1890s, replacing steam engines with electric motors, yet aggregate manufacturing productivity showed no discernible acceleration for roughly three decades afterward. The reason was *architectural* rather than *technological*: early factories simply substituted a single large electric motor for a single large steam engine, preserving the same central-drive layout, the same line-shafts, and the same physical plant design that steam power had required. The productivity gains that electrification ultimately delivered did not arrive until factory owners redesigned their buildings and workflows around distributed unit-drive motors (one small motor per machine) enabling flexible floor layouts, single-story continuous-flow plants, and production sequencing that centralized steam power had made physically impossible. The technology arrived decades before the productivity did, because the complementary reorganization of physical and organizational structure took decades to catch up.

The parallel to the information technology investments of the 1980s and 1990s that Brynjolfsson studied directly is exact enough to have given the phenomenon its name. Economist Robert Solow's well-known observation that computers had become visible everywhere except in the productivity statistics captured a genuine puzzle of the period: firms were investing heavily in computing hardware and software throughout the 1980s, yet national productivity growth remained sluggish. Brynjolfsson's own research, published in the early 1990s, diagnosed the cause in terms that echo David's account of electrification almost precisely… the productivity benefits of information technology were not being generated by the technology investment itself but by the complementary organizational investments, in business process redesign, in decentralized decision rights, in new management practices and skill development, that effective use of the technology required, and those complementary investments lagged the technology purchase by years. When U.S. productivity growth did accelerate meaningfully in the mid-to-late 1990s, the acceleration was concentrated precisely in the firms and industries that had made those organizational investments alongside their technology spending, not simply in the firms that had spent the most on hardware.

This history clarifies the mechanical link between the two curves that AI executives are watching today. The Hype Cycle's Trough of Disillusionment is not an independent, freestanding phenomenon that happens to coincide with the J-curve's value dip. It is, in large part, a lagging and distorted echo of it. Boards, analysts, and the business press observe quarter after quarter of substantial AI investment without a proportional, visible productivity return, and that observation is not wrong… the J-curve's dip is *real*, and the complementary organizational investment that would close it genuinely has not finished. Where sentiment goes wrong is in the size and direction of its correction: because the Hype Cycle is a social and narrative curve rather than a measurement curve, it is prone to overshoot, extrapolating from the accurate observation that gains have not yet materialized to the inaccurate conclusion that the technology has failed to deliver on its promise, rather than the more precise and more useful conclusion that the complementary investment simply has not caught up yet. The trough of collective sentiment, in other words, is frequently a correct diagnosis of a real problem attached to an incorrect prognosis about its cause and its resolution.

The practical diagnostic this distinction makes possible is, for the business leader deciding whether to persist with or retreat from an AI initiative, considerably more useful than either curve read in isolation. The Hype Cycle alone cannot tell a leader whether a specific program is failing or simply early; sentiment about AI broadly is a poor proxy for the health of any particular initiative. The corrective is to look underneath the sentiment reading and ask a small number of concrete questions about whether the complementary organizational investment the J-curve's upward stroke requires is currently underway, rather than treating a falling sentiment index as sufficient grounds for a funding decision on its own:

- Has the AI initiative been accompanied by genuine redesign of the workflows and decision rights it touches, or has AI capability simply been layered on top of processes that remain otherwise unchanged?
- Are leading indicators of organizational capability (e.g., training completion, workflow adoption rates, model usage intensity, data infrastructure modernization) improving even though lagging financial indicators have not yet turned positive?
- Has investment in the infrastructure, governance, and talent that complementary transformation requires continued at a level consistent with a program still building its foundation, or has it been quietly reduced in response to disappointing headlines rather than disappointing evidence from within the program itself?
- And, most pointedly: would abandoning the initiative now represent a considered response to genuine evidence that the underlying transformation has stalled, or would it simply be a response to the discomfort of operating inside a trough that the historical pattern predicted the organization would enter?

A business organization that answers these questions honestly and finds that complementary investment is genuinely progressing has good reason to persist through a falling Hype Cycle sentiment reading, precisely because the two curves are measuring different things and the one that matters for its own program is the one tracking its own organizational transformation, not the one tracking the market's collective mood. An organization that finds complementary investment has stalled (that training has lapsed, that process redesign was never funded past the pilot stage, that the data infrastructure required for AI to act on current rather than historical information was never built) has a genuine problem that sentiment alone would have obscured behind either excessive early enthusiasm or excessive later pessimism. One of the most consequential and most concrete forms that complementary investment takes (the *data architecture* that determines whether AI systems can act on live operational reality rather than a stale historical snapshot of it) is the subject to which this chapter now turns.

Converged Architectures and Real-Time AI: How HTAP Solves the Stale Data Problem

There is a fundamental architectural *contradiction* embedded in the way most enterprise AI systems operate today, and it is one that receives far less attention in AI strategy discussions than its practical significance warrants. The contradiction is this: the business decisions that AI is being deployed to improve are almost universally time-sensitive (fraud is committed in milliseconds, customer churn begins with a single unresolved interaction, supply chain disruptions cascade within hours, and market opportunities open and close in windows that manual decision processes cannot match) yet the AI systems making or informing those decisions are overwhelmingly trained and operated on data that is hours, days, or weeks old by the time it reaches the model.

This is not a consequence of technical negligence or strategic oversight. It is a structural artifact of the data architecture that enterprises have built over the past three decades… an architecture designed around the *separation* of transactional systems that process live operational data and analytical systems that store and query historical data, a separation that made enormous sense in the era of constrained computational resources and that has delivered tremendous value in the form of data warehouses, business intelligence platforms, and the entire modern analytics stack. But that architectural separation, which was a solution to the resource constraints of a prior era, has become a structural liability in the AI era… a design choice made for legitimate historical reasons that now imposes a ceiling on the timeliness, relevance, and ultimately the value of AI-driven decisions across the enterprise. Hybrid Transactional and Analytical Processing, known as HTAP, represents the architectural response to this contradiction: a converged data architecture in which *transactional* and *analytical* workloads operate on the same live data simultaneously, eliminating the extraction, transformation, and loading (ETL) processes that introduce latency between operational reality and analytical insight, and enabling AI systems to act on the world as it actually is rather than as it was when the last data pipeline completed its run.

To fully appreciate why HTAP matters for enterprise AI, it is necessary to understand the architecture it is designed to replace and the specific ways in which that architecture constrains AI value creation. The conventional enterprise data architecture that has dominated since the 1990s is built around a clear and deliberate separation between two distinct categories of system. *Online Transactional Processing* systems (the operational databases that power core business applications including ERP platforms, CRM systems, order management systems, payment processors, and the full range of systems that record and execute the moment-to-moment transactions of the business) are optimized for write-heavy workloads: fast, concurrent recording of individual transactions with the consistency and durability that operational reliability requires. *Online Analytical Processing* systems (the data warehouses, data lakes, and data lakehouses that power business intelligence, reporting, and increasingly AI model training and inference) are optimized for read-heavy workloads: complex queries across large historical datasets that extract patterns, aggregations, and insights from accumulated transactional history.

The bridge between these two worlds is the ETL process… the data pipeline infrastructure that periodically extracts data from operational systems, transforms it into the formats and schemas that analytical systems require, and loads it into the analytical environment where it becomes available for reporting and AI workloads. ETL pipelines are among the most mature and most widely deployed components of enterprise data infrastructure, and they work… within the constraints of the architecture they serve. Those constraints are significant: ETL processes introduce latency that ranges from minutes in near-real-time streaming architectures to hours or days in batch-oriented environments, they require substantial engineering investment to build and maintain, they create data consistency challenges when operational and analytical systems fall out of synchronization, and they produce the analytical environment's defining characteristic from an AI perspective… data that is a historical snapshot of operational reality rather than a live representation of it. For AI systems whose value depends on the currency of the data they act on, this latency is not a minor inconvenience. It is a fundamental constraint on the quality and relevance of every decision the AI system makes.

The practical consequences of stale data in AI systems are more pervasive and more costly than most AI project assessments acknowledge. Consider the fraud detection use case that is among the most widely cited examples of successful enterprise AI deployment. A fraud detection model operating on transaction data that has been processed through a conventional ETL pipeline may be working with data that is anywhere from fifteen minutes to several hours old… a timeframe within which a sophisticated fraud actor can complete dozens of transactions, exhaust account balances, and move funds beyond recovery before the AI system's risk score triggers any intervention. The difference between a fraud detection system operating on fifteen-minute-old data and one operating on live transactional data is not a marginal improvement in detection latency. It is the difference between a system that catches fraud in progress and one that documents fraud after the fact. The same dynamic applies across a remarkably broad range of high-value AI use cases.

A personalization engine recommending products to a customer whose most recent browsing and purchase behavior has not yet propagated through the ETL pipeline is making recommendations based on a customer model that does not reflect the customer's current intent... potentially surfacing products the customer has already purchased, missing the contextual signals that would make a recommendation genuinely relevant, and delivering the kind of slightly-off personalization experience that customers notice and that erodes the trust that makes personalization valuable. A supply chain AI system optimizing inventory allocation based on demand signals that are a day old is operating in a world that no longer exists... missing the real-time signals from point-of-sale systems, logistics tracking platforms, and supplier inventory feeds that would allow it to respond to emerging disruptions before they cascade into stockouts or overstock positions. A dynamic pricing system that cannot incorporate real-time competitor pricing, current inventory levels, and live demand signals into its optimization is not a dynamic pricing system in any meaningful sense. It is a sophisticated historical analysis tool masquerading as a real-time decision engine, and the gap between what it appears to be and what it represents an AI capability promise that the underlying architecture is structurally incapable of keeping.

HTAP architecture addresses this constraint by eliminating the separation between transactional and analytical processing at the data layer, enabling both workloads to operate concurrently on the same live dataset without the ETL intermediary that conventional architecture requires. The technical innovation that makes this possible is the development of database engines and distributed data platforms capable of efficiently handling both the write-intensive, high-concurrency workloads of transactional processing and the read-intensive, complex-query workloads of analytical processing within a single unified system... a capability that was beyond the reach of database technology for most of the history of enterprise computing but that advances in in-memory computing, columnar storage, distributed processing, and hardware architecture have made increasingly practical at enterprise scale. Several distinct technical approaches to HTAP have emerged in the market, representing different points on the spectrum between full convergence and coordinated separation.

Truly converged HTAP systems (exemplified by platforms such as SingleStore, TiDB, and Oracle AI Database 26ai) maintain a single copy of data in a format optimized for both transactional and analytical access simultaneously, using sophisticated storage engine designs that serve both workload types without the consistency trade-offs that earlier attempts at convergence produced. Tightly coupled HTAP architectures (represented by approaches such as Google's AlloyDB, Microsoft's SQL Server with in-memory OLTP, and various NewSQL platforms) maintain separate but tightly synchronized storage layers for transactional and analytical workloads, with replication latency measured in milliseconds rather than the minutes or hours of conventional ETL pipelines. Streaming integration architectures (built on platforms such as Apache Kafka, Apache Flink, and Confluent) do not fully eliminate the separation between transactional and analytical systems but dramatically reduce the latency of data movement between them, enabling near-real-time analytical and AI workloads on data that reflects operational reality within seconds of the underlying transactions occurring. Each of these approaches involves trade-offs in cost, complexity, consistency guarantees, and workload performance that enterprise architects must evaluate against their specific use case requirements, but collectively they represent a maturing ecosystem of technical options for enterprises ready to move beyond the architectural constraints that ETL-dependent data infrastructure imposes.

Oracle AI Database 26ai, which reached general availability in January 2026 as the successor to Oracle Database 23ai, illustrates the direction in which this convergence is moving in practice. Rather than treating AI capability as a layer bolted onto a separate analytical environment, Oracle AI Database 26ai folds AI vector search, in-memory acceleration for mixed transactional and analytical workloads, and support for agentic AI workflows directly into the core database engine, allowing AI inference and retrieval to operate against live operational data without the extraction step that has traditionally separated the systems that run a business from the systems that help decide how to run it better. For enterprises already operating Oracle-based transactional systems, the practical significance is architectural rather than cosmetic: the same platform that records the transaction can, in the same moment, reason over it. This is representative of a broader competitive response across the database vendor landscape rather than a claim that any single platform is the correct choice for a given enterprise, but it is meaningful evidence that the market has moved decisively toward treating real-time, AI-native data access as a baseline expectation for modern DBMS platforms rather than a premium or experimental feature.

The implications of HTAP architecture for AI system design and capability are transformative across several dimensions that deserve careful examination by AI strategists and enterprise architects alike. The most immediate and most obvious implication is inference quality… the improvement in AI decision relevance and accuracy that results from models operating on current rather than historical data. But the deeper and more strategically significant implication is the emergence of a new category of AI application that is simply not possible within conventional ETL-dependent architecture: what practitioners are beginning to call transactional AI… AI systems that are not only informed by transactional data but that participate directly in transactional processes, reading live operational context, generating decisions or recommendations, and writing outcomes back to operational systems within the latency envelope of the transaction itself. A credit decisioning system that can incorporate a customer's transaction behavior from the last thirty seconds, their current account balance in real time, and live fraud signals from the current session into a lending decision that completes within the response time of a mobile banking application is a qualitatively different system from one that scores applications against a customer model updated nightly. A clinical decision support system that can incorporate a patient's vital signs from the last five minutes, their medication administration record updated in real time, and live laboratory results as they are reported into a treatment recommendation that reaches the clinician before the next assessment is a qualitatively different system from one working from data synchronized at the last ETL batch. These are not incremental improvements to existing AI applications. They are new categories of AI value creation that require HTAP or equivalent architectural capability as a prerequisite, and their emergence represents what may be the most significant frontier in enterprise AI architecture over the next decade.

The organizational and strategic implications of HTAP adoption extend beyond the technical architecture into the investment priorities, talent requirements, and governance frameworks that enterprise AI programs must evolve to support real-time AI at scale. The transition from ETL-dependent to HTAP-enabled AI infrastructure is not a simple platform migration. It requires rethinking data modeling approaches that have been optimized for batch processing, redesigning AI inference pipelines that have been built around the assumption of periodic rather than continuous data availability, retraining the data engineering teams whose skills have been built around ETL tooling and batch orchestration frameworks, and establishing new governance practices for AI systems that make decisions in real time without the human review checkpoints that batch processing architectures naturally create. The cost and complexity of this transition should not be underestimated, and enterprises that approach it without a clear strategic framework for prioritizing which AI use cases justify the investment in real-time data infrastructure (and which can continue to operate effectively on near-real-time or batch data) will find the transition more expensive and more disruptive than necessary.

The framework for making these prioritization decisions is fundamentally a question of decision latency value: how much additional business value does a specific AI decision generate if it is made on data that is current to the second rather than current to the hour or the day, and is that incremental value sufficient to justify the architectural investment required to deliver it. For fraud detection, dynamic pricing, real-time personalization, clinical decision support, and algorithmic trading, the answer is clearly and compellingly affirmative. For strategic planning AI, long-range demand forecasting, and HR analytics applications, the answer is far less clear, and the architectural investment in real-time data delivery for these use cases may generate returns that do not justify the cost. The AI-ready enterprise of the next decade will be one that has built the architectural capability to support real-time AI where the decision latency value justifies it, while maintaining the cost efficiency of conventional analytical architecture where it does not... a hybrid infrastructure strategy that requires exactly the kind of nuanced, use-case-driven architectural judgment that distinguishes genuine AI infrastructure strategy from the technology procurement decisions that many organizations mistake for it.

The convergence of HTAP architecture with the broader trajectory of enterprise AI development points toward a future data architecture that looks fundamentally different from the layered, latency-accepting, ETL-dependent infrastructure that currently underpins most enterprise AI deployments. In this emerging architecture, the distinction between operational and analytical systems (the organizing principle of enterprise data infrastructure for three decades) dissolves into a unified data fabric that serves all workload types from a single live representation of operational reality, with AI inference woven into transactional processes rather than layered on top of historical data extracts, and with the feedback loops between AI decisions and operational outcomes closing in real time rather than in the next day's batch cycle. This is not a description of a distant theoretical future. The technical components of this architecture exist *today*, are being deployed by leading enterprises in specific high-value use cases, and are maturing rapidly toward the reliability, scalability, and cost efficiency that enterprise-wide adoption requires. The enterprises that begin building toward this architectural future now (investing in HTAP-capable data platforms, developing the data engineering and AI infrastructure skills that real-time AI requires, and identifying the high-value AI use cases where real-time data access generates the most significant competitive advantage) are not making a speculative bet on an uncertain technology trajectory. They are making a well-evidenced investment in the infrastructure foundation that the next generation of enterprise AI capability will require and positioning themselves to build that capability at a time when the competitive advantage of doing so is still available to be captured.

The Emergence and Evolution of the Chief AI Officer (CAIO)

In the contemporary landscape of digital transformation, the rapid proliferation of generative and predictive artificial intelligence as noted previously has necessitated a new tier of executive leadership: the Chief AI Officer (CAIO). While previous technological shifts were often managed under the broader umbrella of the Chief Technology Officer (CTO) or Chief Information Officer (CIO), the unique ethical, strategic, and operational complexities of AI demand a specialized steward. The CAIO does not merely oversee technology; they serve as the primary architect of an organization's

AI-driven future, balancing the pursuit of aggressive business growth with the stringent requirements of ethical governance and organizational readiness.

Strategic Alignment and Value Creation

The foundational mandate of the CAIO is the development and execution of a comprehensive AI Strategy. Unlike traditional IT strategies that focus on infrastructure and uptime, an AI strategy is intrinsically linked to the organization's value proposition. The CAIO is responsible for defining high-level AI goals that align seamlessly with broader business objectives, ensuring that every pilot program or implementation is designed with measurable business impact in mind. This involves a rigorous process of prioritizing AI initiatives based on potential ROI, scalability, and competitive advantage. By translating abstract machine learning capabilities into concrete business outcomes, the CAIO ensures that AI remains a strategic asset rather than a series of disconnected technical experiments.

The Pillar of Governance and Ethical Stewardship

Beyond strategic growth, the CAIO acts as the guardian of the organization's integrity through AI Governance. As AI systems become more autonomous and influential in decision-making, the risks (ranging from algorithmic bias and data privacy breaches to regulatory non-compliance) grow exponentially. The CAIO establishes robust governance frameworks that ensure the ethical and responsible use of AI across all departments. This role involves managing the lifecycle of AI risks and implementing "Responsible AI" protocols that prioritize transparency and accountability. In this capacity, the CAIO serves as the point of contact for regulators and auditors, bridging the gap between innovative speed and societal safety.

Integration Through Cross-Functional Leadership

The CAIO's effectiveness is largely determined by their ability to exercise Cross-Functional Leadership. AI is rarely a localized technology; its implementation typically ripples through marketing, supply chain, human resources, and finance. Consequently, the CAIO must act as a bridge between the technical engineering teams and the functional business units. This integration requires a leader who can speak the languages of both data science and corporate finance, facilitating collaboration and guiding cross-departmental AI audits. By fostering this connectivity, the CAIO ensures that AI tools are not just deployed but are deeply integrated into the operational fabric of the company, maximizing their utility and adoption.

Distinguishing the C-Suite: CAIO, CTO, and CDO

To fully understand the CAIO role, it must be contextualized alongside the Chief Technology Officer (CTO) and the Chief Data Officer (CDO). While these roles are complementary, their focuses remain distinct. The CTO is traditionally the builder and maintainer of the technical infrastructure… the "pipes" and "hardware" of the organization. The CDO is the curator of the information itself, managing data quality, security, and accessibility… providing the "fuel" for the enterprise. In this ecosystem, the CAIO is the orchestrator who applies that data through advanced algorithms to solve specific business problems. If the CTO builds the engine and the CDO provides the fuel, the CAIO determines the destination and ensures the vehicle is driven safely and efficiently. Together, this "technical triad" forms the backbone of a modern, data-driven enterprise.

Cultivating an AI-Ready Culture

The Chief AI Officer can architect the strategy, install the governance framework, and orchestrate the cross-functional coordination that enterprise AI requires, but none of it converts into durable value unless the organization surrounding that role changes how it works, thinks, and tolerates uncertainty. Culture is the substrate the CAIO cultivates rather than a program the CAIO can unilaterally install, and it deserves examination (in its own right) because it is, in the end, the organizational characteristic that most directly determines whether a business enterprise experiences the J-curve as a temporary and survivable trough or as a permanent stall. An AI-ready culture is built on the same foundation this chapter identified at the outset as characteristic of the AI-ready organization (i.e., leadership fluent enough to interpret probabilistic model output rather than demand false certainty from it, data practices disciplined enough to be trusted, and deployment processes iterative enough to treat failure as information rather than embarrassment) but culture is where those characteristics either take root in daily behavior or remain aspirational language in a strategy document.

Concretely, cultivating this culture means normalizing continuous learning as an operational expectation rather than an occasional training event: workforces at every level, not only technical staff, need structured opportunities to build genuine fluency with the AI tools reshaping their work, delivered with enough candor about present limitations that employees develop calibrated trust rather than either uncritical reliance or reflexive resistance. It means treating experimentation as a legitimate and rewarded activity rather than a deviation to be minimized, because probabilistic systems produce failures as a routine part of their operation, and organizations that punish the team whose pilot did not work discourage exactly the iteration that eventually produces the teams whose pilot does. And it means recognizing that culture is what sustains an organization's investment commitment through the specific period (i.e., the bottom of the J-curve, the trough of the Hype Cycle) when the pressure to retreat is strongest and the evidence that persistence will pay off is least visible. An organization whose culture treats the value dip as confirmation of failure will retreat from it; an organization whose culture has internalized the shape of the transformation journey in advance is far more likely to reach the inflection point on the other side. Culture, in this sense, is not a soft complement to the harder disciplines of strategy, governance, and architecture examined elsewhere in this chapter. It is the mechanism that determines whether a business enterprise completes the journey those disciplines chart or abandons it partway through.

The Future of Enterprise AI Adoption: What the Patterns Tell Us

Strategy, governance, architecture, leadership, and culture do not operate in isolation from one another; they are the interacting components of a single transformation, and it is worth stepping back, now that each has been examined individually, to ask what their combined trajectory suggests about where enterprise AI adoption is in-fact heading.

The accumulated weight of everything this book has examined (namely the technology trajectories, the organizational dynamics, the economic frameworks, the governance architectures, the failure patterns, and the competitive implications of AI adoption) converges on a set of conclusions about the future of enterprise AI that are neither the uncritical optimism of the hype peak nor the reflexive skepticism of the trough, but something more useful than either: a grounded, evidence-based projection of how *AI will reshape* enterprise organization, competitive dynamics, and economic value creation over the next decade, and what the enterprises best positioned to lead that reshaping are doing differently right now. The future of enterprise AI adoption is not a single story. It is several simultaneous stories unfolding at different speeds across different industries, organizational contexts, and technology maturity curves, and the business leaders who will navigate it most successfully are those who can hold that complexity without either collapsing it into false simplicity or being paralyzed by its genuine uncertainty. What follows is not a prediction in the narrow sense of a forecast that will be proven right or wrong by specific events on specific dates. It is a structural projection… an identification of the forces, dynamics, and organizational realities that will shape enterprise AI adoption regardless of which specific technologies emerge, which vendors dominate, or which regulatory frameworks ultimately prevail, because they are rooted in patterns of human organization, economic competition, and institutional behavior that have proven remarkably consistent across prior technology transitions of comparable significance.

The *first and most foundational conclusion* is that enterprise AI adoption is moving, irreversibly and accelerating, from a discretionary strategic initiative to a baseline operational requirement… from something that forward-thinking organizations do to something that all viable organizations must do to remain competitive in their markets. This transition, which the Hype Cycle framework suggests is characteristic of the passage from the Slope of Enlightenment to the Plateau of Productivity, has already occurred in several industry sectors and is underway in virtually all others. In financial services, the question is no longer whether AI will be used for credit decisioning, fraud detection, algorithmic trading, and customer experience personalization, but how well each institution's AI capabilities compare to its competitors' and whether its governance framework is sophisticated enough to satisfy increasingly demanding regulatory expectations. In healthcare, AI-assisted diagnostics, drug discovery, clinical trial design, and care pathway optimization have moved from research novelty to competitive differentiator to, in the leading health systems, operational standard. In retail, supply chain optimization, demand forecasting, personalization engines, and dynamic pricing systems powered by AI have separated the leaders from the laggards with sufficient clarity that the gap is now visible in margin structures and customer retention metrics. The industries where this transition is still in its early stages (i.e., professional services, education, government, and significant portions of manufacturing and construction) are not exceptions to the pattern. They are earlier points on the same curve, and the organizations within them that treat their current distance from the AI frontier as a reason for complacency rather than urgency are making a strategic error whose cost will become apparent at precisely the moment when it is most expensive to correct.

The *second conclusion* is that the primary axis of AI competitive differentiation is shifting from access to capability. In the early years of the current AI wave, competitive advantage in AI was largely a function of access… access to large language models that only a handful of organizations could develop, access to the computational infrastructure required to train and deploy sophisticated AI systems, access to the data science talent that could build and maintain AI applications, and access to the capital required to fund all of the above. That access advantage has been substantially eroded by the commoditization of AI infrastructure through cloud platforms, the democratization of foundational model capability through API access, the rapid expansion of the AI talent supply through university programs and self-directed learning, and the dramatic reduction in the cost of AI development that open-source models and improved tooling have produced. What cannot be commoditized (i.e., what remains genuinely scarce and genuinely differentiating) is the organizational capability to deploy AI effectively: the proprietary data assets that reflect unique operational experience, the workflow architectures that embed AI into business processes in ways that generate compounding rather than static value, the human-AI collaboration practices that extract the full potential of AI systems rather than using them as expensive substitutes for prior manual processes, and the governance infrastructure that allows organizations to move fast without generating the regulatory, reputational, and operational risks that have derailed less disciplined AI programs. The implication for enterprise AI strategy is consequential: investment in AI technology infrastructure, while necessary, is no longer sufficient to generate durable competitive advantage. The *scarce resource* is organizational capability, and the organizations that invest in building it (in training, in process redesign, in governance, in change management, in the human architecture that makes AI adoption genuinely transformative rather than superficially impressive) are the ones that will convert the commoditized technology layer into proprietary competitive advantage.

The *third conclusion* concerns the organizational architecture of the enterprise itself, which AI is in the process of reshaping more fundamentally than any technology since the introduction of the internet, and possibly more fundamentally than that. The traditional enterprise organizational model (i.e., hierarchical, functionally siloed, built around human cognitive bandwidth as the primary constraint on decision-making speed and quality) was designed for a world in which information processing was expensive, expertise was scarce and difficult to distribute, and the coordination costs of complex decisions required organizational structures that concentrated authority at levels where sufficient expertise and information could be assembled to support it. AI systematically removes or dramatically *reduces* each of these constraints simultaneously. When AI systems can process and synthesize information at speeds and scales that human cognition cannot approach, when expertise embedded in models can be distributed to every point in the organization that needs it, and when decision support of extraordinary quality can be made available to frontline employees in real time, the organizational structures designed to manage the constraints that AI removes become not just unnecessary but actively counterproductive… adding coordination cost, slowing decision velocity, and inserting human bottlenecks into processes that AI has made capable of operating faster and more accurately without them.

The enterprises of the *next decade* will increasingly be organized not around the management of human cognitive bandwidth but around the orchestration of human and AI capability in combinations that leverage the distinct strengths of each, namely human judgment, contextual wisdom, ethical reasoning, relational intelligence, and creative synthesis on one side;

AI speed, consistency, pattern recognition across vast data, and tireless availability on the other. This is not a marginal adjustment to existing organizational models. Rather, it is a structural redesign of how enterprises create and coordinate value, and the organizations that begin that redesign intentionally (as opposed to waiting for competitive pressure to force it under less favorable conditions) will define the organizational templates that others eventually follow.

The *fourth conclusion* addresses the human dimension of enterprise AI's future in a way that cuts against the grain of much current AI discourse, which oscillates between anxiety about human displacement and reassurance about human augmentation without fully reckoning with the more complex and more interesting reality that lies between these poles. The future of enterprise AI is not a future in which humans are replaced by machines, nor is it a future in which AI simply makes existing human roles more efficient without changing their fundamental nature. Today, we can see a future in which the nature of human contribution to enterprise value creation is genuinely and substantially transformed… in which the skills, capabilities, and forms of judgment that create economic value shift in ways that advantage some individuals and disadvantage others, that make some organizational roles more central and some less relevant, and that require continuous adaptation from virtually everyone engaged in knowledge work. The enterprises that will navigate this transformation most successfully are those that invest genuinely (not rhetorically) in their people's capacity to evolve with it: building learning organizations that treat continuous capability development as an operational priority rather than an HR function, creating career architectures that provide meaningful pathways for workers whose current roles are being transformed by AI, and cultivating the human capabilities (i.e., critical thinking, ethical judgment, creative synthesis, complex communication, and the ability to ask better questions of AI systems) that will be most valuable precisely because they are most difficult to automate. The human ceiling that this book has described as the binding constraint on AI value realization today will remain the binding constraint on AI value realization in the future, not because the technology will fail to advance but because the organizational and human capability required to use advancing technology well will always be the scarcer and more consequential resource.

The *fifth and final conclusion* is simultaneously the most uncertain and the most important: the enterprises that will lead in the AI era are not yet fully determined, and the window during which the competitive positions of the AI Plateau are being established remains open… but it will not remain open indefinitely. The history of general purpose technology transitions (such as electricity, the internal combustion engine, the computer, and the internet) consistently shows that the period during which new entrants and adaptive incumbents can establish the positions that will define industry leadership for the subsequent generation is real but time-limited, and that the closing of that window is rarely announced in advance. It becomes apparent only in retrospect, when the compounding advantages of the leaders have grown large enough to make catching up prohibitively expensive. The specific timeframe within which this window closes for AI will vary by industry, by technology category, and by the pace at which regulatory frameworks and infrastructure standards mature… but the directional signal from everything the Hype Cycle pattern, the competitive dynamics analysis, and the organizational capability evidence collectively suggest is consistent: the window is measured in years, not decades, and it is *narrowing*. The organizations that treat this moment (this specific, unrepeatable moment in the development of the most consequential general-purpose technology of the century) with the strategic seriousness, investment discipline, organizational commitment, and genuine human respect that it demands are not merely making a good technology decision. They are making a choice about what kind of organization they intend to be in a world that AI is already in the process of remaking, and whether they intend to lead that remaking *or* be remade by it.

Final Thoughts

This chapter opened at the horizon (with the still-uncertain trajectory toward AGI and the strategic, ethical, and organizational demands that even today's narrow AI already places on the enterprise) and it closes, appropriately, back at ground level, with the specific and practical disciplines that determine whether any individual organization actually reaches that horizon in a position of strength. The argument this book has built, chapter by chapter, arrives here at a small number of load-bearing determinations that this final chapter has tried to make concrete rather than aspirational.

The *first* is that two of the most widely cited frameworks for understanding AI's trajectory (i.e., the Gartner Hype Cycle and Brynjolfsson's productivity J-curve) are not redundant confirmations of the same pattern but distinct instruments measuring different things: one a curve of market sentiment rooted in the psychology of technology adoption, the other a curve of measured economic output rooted in the pace of complementary organizational investment. Their resemblance in shape reflects a genuine mechanical relationship (sentiment crashes into its trough substantially because the productivity dip beneath it is real) but that relationship runs through complementary investment, not through the technology itself, and the historical parallels of factory electrification and the 1990s information technology productivity paradox both confirm the same lesson across a full century of general purpose technology adoption: the gains arrive on the organization's schedule, not the technology's.

The *second* determination is that this diagnostic has a concrete architectural expression. The data infrastructure separating transactional and analytical systems (an architecture built for entirely legitimate reasons three decades ago) has become one of the specific complementary investments an enterprise must make to close the gap between AI's potential and its realized value, and the emergence of converged, HTAP-capable platforms represents genuine progress in closing it, not merely a vendor feature race.

The *third* determination is that none of this happens without deliberate leadership and cultural investment. The Chief AI Officer role that has emerged to steward AI strategy, governance, and cross-functional integration is itself evidence that enterprises are recognizing AI transformation as a distinct executive discipline rather than a subset of existing technology management, and the culture that role cultivates (one that tolerates the discomfort of the value dip, rewards disciplined experimentation, and treats continuous learning as an operational necessity) is what ultimately determines whether an organization's AI investment compounds into durable advantage or stalls in the trough that this chapter has described in detail.

None of these determinations offer a shortcut around the value dip, the trough of disillusionment, or the hard work of organizational redesign that both curves this chapter examined identify as the actual source of AI's return. What they offer instead is a *map*… of where the gains genuinely come from, of how to tell a healthy trough from a genuine failure, and of what an organization must build, deliberately and patiently, to be standing on the plateau when the rest of the market arrives there. That map does not make the journey shorter. It makes the journey *navigable*, and for the enterprises willing to walk it with the discipline this book has described, it makes reaching the AI horizon this chapter opened with a matter of organizational *will* rather than technological chance.

GLOSSARY OF TERMS

A

Accountability The principle that organizations—and the individuals who design, deploy, or oversee AI systems—are responsible for the outcomes those systems produce, including harms caused by erroneous or biased predictions.

Active Learning A data-labeling strategy in which a machine-learning model identifies the most informative or uncertain samples and requests human annotation. This reduces labeling cost and accelerates model improvement.

Activity Data (High-Volume / High-Frequency) Operational or user interaction data generated continuously through real-world activity. Becomes a strategic asset when exclusive, large-scale, and collected in feedback loops that strengthen proprietary models.

Activity-Based Cost Attribution (AI Context) A financial method enabled by token or inference billing that maps AI consumption costs directly to processes, products, or business units. Supports transparency and disciplined scaling.

Adversarial Attack A technique that manipulates AI models by subtly altering inputs (e.g., adding noise or placing stickers on stop signs) to force incorrect predictions. Represents a critical security risk for deployed AI systems.

Adversarial Examples Inputs intentionally crafted to deceive an AI model—often by making small, imperceptible changes—to cause incorrect predictions. These represent a significant security risk for deployed systems.

Adversarial Robustness The resilience of AI models against malicious inputs designed to manipulate predictions. Enhanced through techniques like adversarial training, anomaly detection, and robustness testing.

Agent Team (Autonomous Systems) A coordinated group of AI agents that collaborate to perform complex tasks, requiring human supervision, orchestration, and exception management at scale.

AGI Pathways Three dominant approaches to pursuing artificial general intelligence: (1) scaling deep learning architectures, (2) integrating symbolic reasoning with probabilistic learning, and (3) orchestrating multi-agent systems to solve open-ended tasks.

AI Center of Excellence (AI CoE) A centralized group responsible for governance, best practices, infrastructure, and advanced modeling expertise, supporting federated AI teams across business units.

AI Ethics Framework A set of guiding principles—transparency, accountability, human rights, fairness—adapted from organizations such as OECD, IEEE, or the EU and operationalized within enterprise policy.

AI Fluency (Executive) The leadership capability to interpret probabilistic model outputs, evaluate risks, and align AI opportunities with business strategy.

AI Governance The set of organizational structures, policies, and processes ensuring that AI systems are developed and deployed responsibly, legally, and ethically. Includes data oversight, model governance, risk controls, and regulatory compliance.

AI Governance Committee A cross-functional oversight body that reviews AI initiatives for ethical alignment, regulatory compliance, risk posture, and business justification before deployment.

AI Life Cycle (Scoping–Data–Modeling–Deployment) The iterative sequence of phases through which AI systems are designed, built, deployed, monitored, and refined. Emphasizes continuous improvement rather than linear completion.

AI Literacy (Workforce) The baseline understanding employees need to responsibly interact with AI systems—covering model limitations, data dependencies, bias risks, and interpretability.

AI Maturity Model A structured framework for assessing an organization's readiness to scale AI. Evaluates capabilities across data quality, infrastructure, governance, talent, cultural preparedness, and operational practices. Guides sequencing of investments and identifies bottlenecks.

AI Moat A durable competitive advantage created through proprietary data, unique models, vertical integration, ecosystem positioning, or platform dominance. Designed to make AI capabilities difficult or costly for competitors to replicate.

AI Moat Diagnostic Checklist A five-question self-assessment used to gauge the current strength of an organization's AI-driven competitive advantage across distinct moat dimensions. Affirmative, evidence-based answers (not merely aspirational ones) to at least two questions signal a genuinely defensible position.

AI-Native Transformation A form of transformation in which processes, products, and business models are redesigned from the ground up using predictive, generative, or autonomous AI capabilities. Goes beyond digitization by reimagining what the organization could be, not just optimizing what it already does.

AI Operating Model The structural design that determines how AI capabilities are organized and scaled—centralized, federated, or hybrid—specifying roles, responsibilities, governance, and standards.

AI Orchestrator An emerging role responsible for managing interactions among multiple AI agents, aligning autonomous system behavior with business objectives, and ensuring safety and accountability.

AI Portfolio A curated set of AI initiatives balanced across short-term, incremental improvements and long-term, transformative bets. Managed similarly to an investment portfolio—evaluated by impact, feasibility, risk, resource requirements, and strategic alignment.

AI-Ready Organization An enterprise that combines the right talent, culture, data governance, operating model, and MLOps capabilities to adopt AI at scale.

AI Review Board (ARB) A multidisciplinary committee—typically including legal, data science, ethics, security, and business stakeholders—that evaluates proposed AI use cases for ethical, technical, and regulatory risks before deployment.

AI Strategy A business-led, enterprise-wide plan that uses AI to achieve measurable financial and operational outcomes. Distinct from a technical roadmap, it defines priorities, value levers, investment sequencing, talent needs, governance, and long-term transformation pathways.

AI Talent Pipeline The set of strategies—university partnerships, global recruiting, internal training, and reskilling—used to attract and develop specialized AI expertise.

AI Translator (Business Liaison / Product Owner) A critical role bridging technical teams and business stakeholders. Defines business problems, aligns KPIs, interprets model outputs, and drives adoption and change management.

AI Winter A period of reduced funding, slowed research progress, and declining public enthusiasm for AI, typically caused by unmet expectations and technological limitations. The major AI winters occurred between the 1980s and early 2000s.

AIOps (AI for IT Operations) A practice that uses AI, ML, and analytics to automate, optimize, and manage IT systems. Involves real-time anomaly detection, event correlation, predictive maintenance, and automated remediation.

Algorithm A mathematical model or set of computational procedures used by AI systems to analyze data, identify patterns, make predictions, or generate decisions. Algorithms range from simple statistical methods to complex deep learning architectures.

Algorithmic Bias Systematic and unfair discrimination produced by an AI model, often originating from biased or unrepresentative training data. Can manifest as unequal error rates, skewed decision patterns, or amplified historical inequities.

Algorithmic Feedback Loop A self-reinforcing cycle in which biased model predictions influence real-world behavior, generating new data that deepens the original bias. Without intervention, such loops institutionalize inequity.

Algorithmic Impact Assessment A structured evaluation, paired with model cards and system documentation, that records an AI system's business purpose, architecture, training data, evaluation results, and known limitations, supporting the transparency and accountability regulators and auditors increasingly expect for high-stakes deployments.

Algorithmic Trading A reinforcement-learning or rules-driven approach to executing trades automatically by analyzing market conditions and optimizing decision sequences to maximize returns.

Anomaly Detection (Fraud & Operations) Models that identify deviations from normal patterns—used extensively in fraud detection, cybersecurity, and operational risk management.

Anonymization A process for removing personally identifiable information (PII) so that individuals cannot be reidentified. Often used to meet privacy regulations and protect sensitive data used in training.

API (Application Programming Interface) A standardized interface that allows software components—including data pipelines, model services, and retrieval systems—to communicate and exchange data programmatically.

Artificial Intelligence (AI) Technology designed to perform tasks that normally require human intelligence—such as learning, reasoning, pattern recognition, and decision-making—with the objective of generating measurable business value.

Auditability The ability to trace and verify how an AI system was trained, deployed, and operated—including data sources, model versions, assumptions, performance metrics, and decision logic.

Augmentation (Human–AI Collaboration) A strategy in which AI supports and enhances human work by handling data-heavy or repetitive tasks, allowing people to focus on creative, strategic, and judgment-based activities.

Automated Feature Extraction A capability of deep learning in which neural networks learn relevant data representations (edges, textures, objects, linguistic patterns, etc.) directly from raw input without manual feature engineering.

Automation (Traditional) The execution of predefined, rule-based tasks without variation. Unlike AI, traditional automation does not learn or adapt; it performs the same actions consistently according to programmed instructions.

Automation Uplift The measurable cost savings or efficiency improvements produced when AI automates routine tasks, such as document processing, customer service triage, or quality checks.

Autonomy Data Advantage A specialized subset of proprietary data used in training autonomous systems (e.g., sensor logs, fleet telemetry). Creates significant barriers to entry because replication requires years of real-world collection.

B

Balanced Scorecard for AI A multi-criteria evaluation approach that integrates financial, operational, customer, and ethical metrics to assess AI performance beyond ROI alone.

Barrier to Entry (AI Context) A structural challenge that prevents competitors from quickly matching an organization's AI capabilities. Often arises from exclusive datasets, tight integration, or entrenched ecosystem effects.

Batch Processing A data-processing method where large volumes of information are collected and processed at scheduled intervals. Historically dominant, now supplemented or replaced by streaming-first architectures.

Batch vs. Real-Time Inference Two deployment modalities: periodic batch scoring for large datasets, and low-latency real-time predictions for interactive or transactional use cases.

Behavioral Signals (Marketing & Sales) Features such as website interactions, trial usage, or email activity used in ML-driven lead scoring and personalized engagement.

Bias Mitigation Techniques for reducing discriminatory outcomes in AI models, including data re-weighting, fairness constraints, prediction auditing, and adversarial fairness testing.

Big Data Extremely large and complex datasets—generated from digital activity, sensors, transactions, and more—that enable modern AI systems to learn patterns and improve performance.

Black Box Model A model whose internal decision-making processes are opaque or difficult to interpret. Common in deep learning, creating challenges for explainability and regulatory scrutiny.

Blue–Green Deployment A release strategy in which a new model or service is launched alongside the existing version, allowing traffic to be switched gradually and safely between the two.

Build-Buy-Partner Decision A strategic framework for determining how AI capabilities should be acquired. Build: maximum control and differentiation; high talent/infrastructure investment. Buy: fastest deployment; limited customization. Partner: hybrid approach leveraging internal expertise and external specialization.

Business-Driven Model Selection A decision-making process that prioritizes financial and operational impact—false positives vs. false negatives—over purely technical accuracy metrics.

Business Model Innovation (AI-Enabled) New revenue models, pricing structures, or value propositions made possible by AI. Examples include outcome-based pricing, personalized subscription tiers, or predictive service offerings.

Business Value (AI) The measurable benefit an organization gains through AI, typically expressed through cost reduction, revenue generation, or risk mitigation.

C

Canary Release A controlled deployment strategy in which a small portion of production traffic is routed to a new model to validate behavior before full rollout.

Candidate Matching Algorithms AI systems used in HR to evaluate resumes, infer job fit, and prioritize applicants based on predicted performance and skill alignment.

Causal + Symbolic Hybrid Models Architectures that combine statistical learning with reasoning and structured knowledge to improve generalization and explainability.

Change Management (AI Adoption) Structured efforts to help employees understand, trust, and effectively use AI tools. Includes training, communication, workflow redesign, and cultural enablement.

Change Management for AI The structured approach to guiding employees through AI-driven workflow changes, including communication strategies, retraining programs, incentives, and phased rollouts.

Change Resistance (AI Adoption) Human reluctance to trust probabilistic predictions, requiring cultural interventions, transparency, and pilot-based learning cycles.

Chief AI Officer (CAIO) The executive role that has emerged as generative and predictive AI have proliferated, distinct from the CTO or CIO, responsible for AI strategy, governance, and cross-functional integration, and for balancing aggressive growth ambitions against ethical and operational readiness requirements.

Cloud Computing On-demand access to scalable computing resources (e.g., servers, GPUs) delivered over the internet, enabling businesses to train, deploy, and operate AI systems without owning physical infrastructure. Also: A model for accessing scalable, on-demand compute, storage, and networking resources over the internet, enabling organizations to train, deploy, and operate AI systems without maintaining physical hardware.

Competitive Disruption (AI-Driven) A market shift caused by firms that use AI to outperform incumbents on pricing, demand prediction, customer experience, or operational efficiency.

Compliance Automation AI-driven tools that monitor regulatory requirements, flag anomalies, generate mandatory reports, and reduce the operational burden of compliance functions.

Compliance-by-Design Embedding regulatory requirements—including documentation, risk controls, and transparency—directly into model development and deployment pipelines.

Compliance Risk The legal exposure arising when AI systems violate data protection laws, fairness requirements, or industry-specific regulations (e.g., GDPR, CCPA, HIPAA).

Computation (Computing Power) The hardware and processing capability required to train and run AI models. High-performance resources such as GPUs or TPUs enable efficient model development and real-time inference.

Compute Orchestration (LLMOps) The management of high-intensity distributed compute—multi-GPU/TPU clusters—for training and serving large language models at enterprise scale.

Computer Vision (CV) A field of AI focused on enabling machines to interpret images, video, and visual sensor data. Common tasks include object detection, image classification, segmentation, anomaly detection, and action recognition.

Concentration Principle The strategic practice of focusing resources on a small number of high-value, high-return AI opportunities instead of pursuing numerous low-impact pilots. Prevents fragmentation and accelerates economic value creation.

Continuous Monitoring (Ethical & Performance) The practice of tracking model behavior over time to detect drift, emerging bias, security vulnerabilities, or regulatory deviations, supported by alerts, dashboards, and periodic audits.

Convolutional Neural Network (CNN) A deep learning architecture optimized for image and video processing. CNNs detect local patterns (edges, shapes, textures) and combine them hierarchically to recognize objects, scenes, and actions.

Cost-of-Error Framework A financial model assessing the relative business impact of false positives and false negatives, guiding model selection and thresholding.

Cross-Functional Data Integration The practice of connecting function-specific datasets—marketing, supply chain, finance, HR—so insights from one area improve decisions in another.

Cross-Functional Discovery A collaborative process in which executives, technologists, operations teams, and compliance stakeholders jointly diagnose pain points and identify AI-enabled solutions. Ensures initiatives are both strategically valuable and operationally feasible.

Cultural Adoption (AI-Ready Culture) An organizational environment that embraces experimentation, probabilistic decision-making, and evidence-based learning, enabling AI to scale beyond isolated pilots.

Customer Lifetime Value (CLV) Optimization (AI-Assisted) The use of predictive analytics to tailor offers, pricing, and engagement strategies to maximize long-term profitability per customer.

D

Data The foundational input that AI systems learn from. High-quality data—characterized by volume, velocity, variety, and veracity—is essential for accurate model performance.

Data Catalog A system of record that documents what data assets an organization possesses, where they reside, how they were collected, what they contain, who is responsible for them, and how they may be used—the organizational infrastructure that makes large-scale data environments navigable and governable.

Data-Centric AI An approach emphasizing systematic improvement of data and features (rather than switching model architectures) to achieve higher performance and reliability.

Data-Centric Culture An organizational mindset that treats data as a strategic asset, with shared responsibility for quality, governance, lineage, and ethical use.

Data Contracts Formal agreements between data producers and downstream consumers that specify schema, semantics, and quality expectations for a dataset, designed to prevent the common failure pattern in which an upstream change—a renamed field, an altered aggregation grain—silently breaks a production AI system.

Data Flywheel A feedback mechanism where more users generate more data, which improves models, which improves product quality, which attracts more users—creating compounding advantage.

Data Governance The organizational framework that assigns data ownership, enforces access controls, regulates usage, and ensures compliance with privacy, security, and ethical standards.

Data Governance Lifecycle The sequence of disciplined stages—beginning with data discovery and cataloging and proceeding through classification, stewardship, and monitoring—by which an organization makes its data assets known, accountable, and fit for both operational and AI use.

Data Integration (Heterogeneous Sources) The process of unifying logs, sensors, ERP systems, customer interactions, and external feeds into coherent datasets for modeling.

Data Labeling (Operational Context) The annotation of historical data to define ground truth—critical in predictive maintenance, fraud detection, or HR attrition models.

Data Lineage A record of where data originated, how it has been transformed, where it is stored, and how it flows through pipelines—critical for debugging, auditing, and regulatory compliance.

Data Lineage (Ethical Context) A record of where training data originated, how it has been altered, and how it flows through the AI system. Critical for transparency, debugging, audits, and regulatory compliance.

Data Mesh A decentralized approach to enterprise data architecture that assigns data ownership to the business domains that generate it rather than to a central platform team, weighed in the book against centralized data platforms on dimensions of efficiency versus domain agility.

Data Moat A defensible strategic advantage created by exclusive, large-scale, high-quality, and difficult-to-replicate datasets. Often the strongest and most enduring form of AI moat.

Data Observability A set of monitoring practices that track data quality, freshness, drift, schema stability, and lineage to ensure reliable inputs for AI systems.

Data Pipeline An orchestrated sequence of processes that ingest, transform, validate, and prepare data for downstream analytics or machine learning. Pipelines ensure reproducibility and consistency between training and production environments.

Data Quality A measure of whether data is accurate, complete, timely, consistent, and representative. High data quality is a prerequisite for trustworthy AI outcomes.

Data Quality Remediation Actions taken to fix missing values, misaligned timestamps, calibration inconsistencies, and corrupted logs—often the longest phase of AI projects.

Data Residency The set of regulatory and architectural constraints governing the physical or jurisdictional location where data may be stored and processed, a consideration that shapes cloud and AI infrastructure decisions for organizations operating across borders.

Data Sovereignty The principle, reinforced by frameworks such as the GDPR and its international counterparts, that data generated within a jurisdiction remains subject to that jurisdiction's laws regardless of where it is technically stored or processed, complicating the architecture of globally distributed AI platforms.

Deep Learning A subset of machine learning that uses multi-layered neural networks to recognize complex patterns in data, powering applications such as image recognition, language processing, and recommendation systems. Also: A subset of machine learning built on multi-layered neural networks. Deep learning excels at processing complex, high-dimensional data such as images, speech, and natural language, but requires large datasets and significant compute resources.

Demand Forecasting (AI-Enhanced) Machine learning–driven predictions of customer demand that inform inventory planning, workforce scheduling, marketing spend, and supplier coordination.

Demographic Parity A fairness metric requiring that different demographic groups receive positive outcomes at similar rates, regardless of underlying characteristics.

Deterministic vs. Probabilistic Systems A distinction between traditional software that follows fixed rules and AI systems that generate probability-based predictions requiring interpretation.

Differential Privacy A privacy-preserving technique that introduces statistical noise to datasets or query results to protect individual identities while preserving aggregate insights.

Digital Transformation Efforts to automate or streamline existing processes through software, workflow tools, or digital channels. Sets foundational capabilities but does not fundamentally reinvent business models the way AI-native transformation can.

Distributed Compute Cluster A network of interconnected compute nodes—often provided by cloud platforms—that work together to train AI models significantly faster than a single machine.

Distributed Computing A model in which computation is spread across multiple machines or nodes to increase scalability and training throughput for large AI workloads.

Documentation (Model Cards / System Cards) Structured records describing a model's purpose, training data, assumptions, metrics, limitations, and monitoring plans. Ensures future teams understand how a model should—and should not—be used.

Documentation Protocols (AI Lifecycle Records) Comprehensive records covering model purpose, data sources, training methods, evaluation metrics, risk assessments, and monitoring plans used for auditing and compliance.

Drift Detection Monitoring changes in data distributions or model performance over time to trigger retraining and recalibration.

Dynamic Pricing Engine Real-time algorithms that adjust prices based on demand, inventory, user behavior, and market conditions. Improves revenue optimization and margin performance.

E

Economic Value (AI Context) The measurable financial benefit generated by AI, such as increased revenue, margin expansion, cost reduction, or risk mitigation. Serves as the primary benchmark for assessing strategic success.

Ecosystem Dominance A competitive position in which a firm orchestrates a network of developers, partners, and users whose participation increases the value of the core platform. Creates reinforcing network effects that are hard for rivals to dislodge.

Ecosystem Leverage The use of external developers or partners to extend the capabilities and value of an AI platform, increasing data flow, stickiness, and innovation.

Edge Computing The practice of deploying models on devices or infrastructure close to where data is generated—such as mobile devices, IoT sensors, or on-premise hardware—to reduce latency and improve privacy.

Embedding A high-dimensional numerical representation of data (e.g., text, images, audio) used to capture semantic meaning. Embeddings are fundamental to vector databases and RAG systems.

Enterprise-Wide AI Platform A unified architecture enabling consistent data access, model deployment, monitoring, and governance across all functions of the business.

Enterprise Recommendation Systems Models that personalize content, product suggestions, or next-best actions using behavioral and contextual data.

Equal Opportunity A fairness standard requiring that error rates (often false negatives) be consistent across demographic groups. Common when equitable access to services (e.g., medical risk assessments) is critical.

ETL / ELT (Extract–Transform–Load / Extract–Load–Transform) Data integration workflows used to move data from source systems to analytics or ML environments. Modern architectures increasingly rely on ELT to leverage in-warehouse compute.

Ethical Drift The gradual misalignment of AI systems with ethical or societal norms as data distributions shift or models evolve, necessitating ongoing monitoring and governance.

Event-Driven Architecture A system design in which data flows are triggered by real-time events (messages, logs, transactions), enabling continuous processing and low-latency AI applications.

Executive Ownership (AI Strategy) The principle that AI strategy must be driven and governed by senior leadership—not isolated within technical teams. Ensures alignment with corporate priorities and sustained resource commitment.

Explainable AI (XAI) Tools and techniques that make model decisions interpretable. Helps developers, auditors, and regulators understand which features influenced a prediction and why. Also: Methods that reveal how models generate predictions, supporting trust, regulatory adherence, and effective human oversight.

F

Fairness Audit A structured evaluation of model outputs across demographic segments to identify disparate impacts and validate equitable performance.

Fairness Metric A quantitative measure used to assess whether a model treats different groups equitably. No single metric applies universally; organizations must choose metrics aligned with context and risk.

Fast Follower Advantage A strategic position achieved by firms that adopt AI slightly later but execute better—avoiding early mistakes, using more advanced technologies, or scaling faster than first movers.

Feature Engineering The process of transforming raw data into meaningful variables that improve model performance. Includes simple transformations and advanced representations like embeddings. Also: Designing meaningful variables that capture patterns—such as vibration amplitudes or engagement frequencies—which often determine model success more than algorithm choice.

Feature Store A centralized repository in ML infrastructure that manages, version-controls, and serves machine learning features reliably for both training and real-time inference. Also: A centralized platform for managing the end-to-end lifecycle of ML features—including definition, documentation, versioning, and serving for both training and inference.

Federated AI Operating Model The "hub-and-spoke" structure balancing centralized governance with embedded domain teams, commonly used in large-scale enterprises.

Federated Learning A privacy-preserving training method where models are trained across distributed devices without transferring raw data to a central location.

Feedback Loop (AI Performance) A cycle in which model outputs influence user behavior and operational outcomes, generating new data that improves model accuracy and competitive position.

Feedback Loop (Algorithmic) See: Algorithmic Feedback Loop.

FinOps (for AI) The financial management discipline adapted to the variable, consumption-based cost structure of AI workloads, addressing the unpredictable compute intensity, training run duration, and inference throughput that distinguish AI infrastructure spending from traditional IT budgeting.

First-Mover Advantage (AI Context) The competitive benefit gained by early entrants who accumulate unique data, define user expectations, and establish standards before rivals can catch up.

Forecasting Models (Demand, Supply, Risk) Predictive systems incorporating historical data, seasonality, promotions, weather, and external factors to support operational planning.

Functional Domain Applications Use cases across marketing, sales, operations, supply chain, finance, and HR that demonstrate AI's practical impact.

G

Gartner Hype Cycle A model describing the pattern by which a new technology moves through inflated expectations, a trough of disillusionment, and a gradual climb to a plateau of productivity; used in the book to explain sentiment about AI as distinct from—but mechanically linked to—its real productivity trajectory.

GDPR / CCPA / HIPAA Key regulatory frameworks governing data privacy, consumer rights, and health information protection. AI systems must comply with these regimes when handling sensitive data.

General AI (Strong AI) A theoretical form of AI that would match or exceed human cognitive abilities across all tasks. It does not currently exist and plays no role in practical business applications.

Generative Model A model capable of producing synthetic data—text, images, sensor traces, or tabular samples—that expands or complements real datasets while reducing privacy or scarcity concerns.

Generative Modeling Deep learning techniques that generate new content—text, images, audio, code—based on learned patterns. Examples include language models and generative adversarial networks.

Global AI Regulatory Landscape The mosaic of AI laws and policies across jurisdictions—EU AI Act, U.S. sectoral guidance, China's algorithm registry—requiring coordinated compliance strategies.

Governance Framework A formal structure defining policies, roles, responsibilities, and processes that ensure AI systems operate ethically, securely, and in regulatory compliance.

GPU (Graphics Processing Unit) A hardware accelerator used to train deep learning models efficiently by performing large-scale, parallel mathematical operations.

H

Hallucination (LLM Context) A phenomenon in which a large language model produces information that appears plausible but is factually incorrect. A critical reliability and safety concern in LLMOps.

High-Risk AI System (EU AI Act) Applications involving safety, credit scoring, employment decisions, or biometric identification that face stringent documentation, testing, and monitoring obligations.

High-Value Use Case An AI opportunity with outsized return potential, typically characterized by data richness, decision intensity, operational significance, and clear financial measurability. Forms the backbone of effective AI strategy.

Historical Bias Bias inherited from the patterns and inequalities embedded in historical data. Models trained on such data may replicate or amplify existing discrimination. Also: Systemic inequality embedded in training data, reflecting past societal patterns that can lead to discriminatory model outcomes if unaddressed.

Homomorphic Encryption An encryption technique allowing computations to be performed on encrypted data without first decrypting it. Useful for privacy-preserving machine-learning workflows.

HTAP (Hybrid Transactional/Analytical Processing) A converged database architecture that allows transactional and analytical workloads to run against the same live data, addressed in the book as the emerging solution to the "stale data problem" created by traditional architectures that separate operational systems from the analytical systems AI models are trained and run on.

Human-in-the-Loop Oversight The requirement that humans remain involved in decision-making processes—especially in high-stakes contexts—to validate outputs, correct errors, and uphold accountability.

I

Industry-Specific AI Disruption The patterns through which AI reshapes sectors differently—such as algorithmic trading in finance, AI triage in healthcare, or autonomous logistics in transportation. Modeled by the distinct data and operational structures of each industry.

Inference (Model Serving) The process of deploying a trained AI model so it can generate predictions in real time or in batch jobs within business operations.

Inference-Based Pricing A usage-based billing model in which organizations pay per inference call or model output instead of fixed licensing fees.

Infrastructure Orchestration Systems (e.g., Kubernetes, Airflow, Dagster) that manage, schedule, and coordinate compute resources, pipelines, and ML workflows, ensuring reliability and observability.

Infrastructure Readiness The maturity of an organization's data pipelines, storage systems, compute environments, model registries, monitoring tools, and governance architecture. Determines the organization's ability to build, deploy, and maintain AI at scale.

Integration Moat A defensive advantage created when AI capabilities are deeply embedded into daily workflows, infrastructure, or physical systems such that switching providers would be operationally or economically prohibitive.

Intelligent Automation Automation enhanced by AI, allowing systems to learn from data, respond to new information, and make context-aware decisions. Unlike traditional automation, it operates effectively under uncertainty.

K

Key Performance Indicator (KPI) — AI Strategy A measurable outcome tied to business objectives—such as revenue uplift, churn reduction, cost savings, or risk reduction—used to evaluate the effectiveness of AI initiatives. Goes beyond technical metrics such as accuracy or latency.

L

Labeling (Annotation) The human or model-assisted process of assigning ground-truth labels to training data. Includes guideline design, quality control, reviewer workflows, and agreement measurement.

Language Processing Unit (LPU) A newer class of AI accelerator, pioneered by Groq, that stores all model weights in on-chip SRAM to eliminate off-chip memory access, producing fully deterministic execution and markedly lower inference latency for large language models.

Large Language Model (LLM) A transformer-based NLP model trained on massive text corpora that can understand, generate, summarize, and classify language, supporting tasks such as chatbots, document analysis, and content creation.

Lead Scoring Models Predictive systems that rank prospects by likelihood to convert, improving sales prioritization and efficiency.

Learning Path Personalization (HR/L&D) AI systems that tailor training content and career development programs to individual employee needs, skill gaps, and performance profiles.

LLMOps (Large Language Model Operations) Operational discipline focused on managing the lifecycle of large language models. Includes prompt management, fine-tuning governance, alignment, safety testing, monitoring, cost optimization, and responsible deployment. Also: Operational practices for training, deploying, and monitoring large language models, including token usage governance, RAG orchestration, and cost monitoring.

M

Machine Learning (ML) A method within AI where algorithms learn from data instead of being explicitly programmed with rules. The system improves its performance as it processes more examples. Also: A family of computational techniques that enable systems to learn patterns and make predictions from data rather than relying on explicit rules. ML powers forecasting, classification, scoring, recommendation, and decision automation.

Machine Learning Operations (MLOps) A set of practices and tooling for deploying, monitoring, versioning, and updating AI models in production to ensure reliability, scalability, and continuous improvement. Also: The discipline of automating model deployment, monitoring, retraining, and CI/CD processes to ensure reliable production performance. Also: A discipline that integrates engineering, data management, deployment, monitoring, and governance to maintain reliable and scalable machine-learning systems in production.

Managed AI Services Cloud-provided tools—including pre-trained models, vector databases, AutoML systems, and end-to-end ML pipelines—that accelerate AI development without requiring organizations to build infrastructure from scratch.

Market Standards (AI-Defined) Protocols, data schemas, interoperability rules, and operational norms established by early movers or platform leaders that influence how an entire industry adopts AI.

Model Compression Techniques such as pruning, quantization, and distillation used to reduce the computational footprint of models while maintaining acceptable accuracy.

Model Degradation The decline in predictive accuracy over time caused by data drift, market shifts, or operational changes. Requires continuous retraining and monitoring.

Model Drift A degradation in model performance caused by changes in underlying data, user behavior, or environmental conditions. Also: A decline in model performance, fairness, or calibration due to changes in underlying data or behavior. Continuous monitoring is required to detect and correct drift.

Model Evaluation (Business-Aligned) Assessing models using financial trade-offs, operational sensitivity, and risk impact—not just precision/recall.

Model Registry A system that tracks model versions, metadata, artifacts, and performance metrics—enabling safe promotion, comparison, rollback, and governance across the ML lifecycle.

Model Risk The risk that a model produces incorrect predictions—due to design flaws, bias, data drift, or implementation errors—which can cause financial, operational, or safety-related harm.

Model Risk Management (MRM) A framework for validating, stress-testing, and monitoring AI models to ensure reliability, stability, and regulatory compliance throughout their lifecycle.

Model Serving (Inference Infrastructure) The specialized systems that host trained models and handle real-time or batch predictions, optimized for scalability, latency, and cost efficiency.

Model Stewardship The ongoing responsibility to maintain, monitor, retrain, and govern AI systems after deployment.

Modern AI Era The current wave of AI innovation beginning in the 2010s, driven by breakthroughs in machine learning, deep learning, big data availability, and cloud computing.

Modern Data Stack A modular, cloud-oriented ecosystem of tools that includes real-time ingestion, scalable storage, distributed compute engines, ML infrastructure, and orchestration frameworks.

Monte Carlo Simulation (AI Valuation) A probabilistic modeling method used by finance teams to evaluate the distribution of possible outcomes for AI initiatives under uncertainty.

Multi-Agent Systems (Enterprise Context) Collections of specialized AI agents collaborating to accomplish complex tasks—coordinated by AI orchestrators or agent supervisors.

Multidisciplinary Review A governance approach that incorporates diverse perspectives (legal, ethical, technical, operational) into decisions about AI deployment, reducing blind spots and ensuring accountability.

Multimodal Data Information expressed in multiple formats—text, images, audio, logs, sensor streams—requiring flexible storage and processing architectures.

N

Narrow AI (Weak AI) AI systems designed to perform a specific task or set of tasks extremely well. They are highly specialized, efficient, and foundational to nearly all current business AI applications.

Natural Language Processing (NLP) A field of AI enabling machines to understand, analyze, generate, and interact using human language across tasks such as sentiment analysis, summarization, translation, search, and conversational interfaces.

Natural Language Processing (NLP) Engagement Tools Chatbots, digital assistants, and conversational AI systems used to automate customer interactions, triage support issues, or augment sales engagement.

Network Effects (AI Ecosystems) The phenomenon in which the value of a platform increases as more users, partners, or developers participate—enhancing data, model performance, and stickiness.

Neural Network A computational architecture composed of interconnected nodes ("neurons") that transform input data through layers of mathematical operations to detect complex patterns.

Normalization (Data Preparation) Standardizing numerical inputs (e.g., sensor values) to ensure stable model performance.

NPU (Neural Processing Unit) A hardware accelerator optimized for neural network workloads, offering higher energy efficiency for AI tasks compared to traditional CPUs or GPUs.

O

Object Detection A computer vision task that identifies which objects appear in an image or video and pinpoints their locations using bounding boxes or segmentation masks.

Operational Embedding The extent to which AI models are integrated into mission-critical workflows. High embedding raises switching costs and creates de facto lock-in.

Operational Integration (Last-Mile AI) Embedding model outputs directly into business processes, dashboards, ticketing systems, and frontline workflows.

Orchestration Framework Tools that manage the sequencing, dependencies, error handling, and monitoring of complex data or ML workflows.

Org Design for AI (2026–2030) Emerging structures emphasizing federated teams, agent supervisors, AI orchestrators, and hybrid human–agent workflows.

Organizational Alignment The harmonization of resources, incentives, culture, governance, and business priorities around AI-driven transformation. Essential for avoiding fragmented or duplicative efforts.

Outcome-Based Business Model A revenue structure in which customers pay based on realized outcomes—uptime, efficiency gains, cost savings—rather than product ownership. Often enabled by predictive analytics.

Outcome-Based Pricing A pricing model enabled by predictive analytics in which customers pay for business outcomes (e.g., guaranteed uptime or performance) rather than products or inputs. Reflects AI's ability to manage risk and forecast results.

P

Parallel Processing Hardware or software capability that allows many computations to occur simultaneously—a critical function for training deep learning models efficiently.

Personalization Engines (Marketing) Models that tailor experience, content, or product recommendations at the individual customer level.

Pilot-to-Scale Framework An adoption model where teams validate early model performance through small pilots, collect feedback, and scale systematically across the enterprise.

Platform Layer (AI Platform) A foundational set of tools, data services, APIs, or models on top of which other companies build. Controlling the platform layer gives firms leverage over an ecosystem and often produces long-term defensibility.

Platform Moat A strategic advantage derived from owning the platform that others depend on for distribution, data, or development. Strengthens over time as more participants join the ecosystem.

Portfolio Management (AI Innovation) The strategic practice of prioritizing AI initiatives based on differentiation potential, long-term defensibility, and alignment with the firm's competitive positioning.

Portfolio Prioritization The process of ranking AI initiatives based on potential impact, feasibility, risk, time to value, and alignment with strategic objectives. Ensures optimal allocation of capital, talent, and technical resources.

Predictive Analytics (Strategic Impact) AI techniques that generate forward-looking insights to optimize key business decisions. Enables new business models, operational efficiencies, and personalized customer experiences.

Predictive Maintenance An AI-enabled approach that forecasts equipment failures before they occur by analyzing sensor data and usage patterns, reducing downtime and maintenance costs. Also: AI systems that detect equipment failure risk, recommend service intervals, and reduce downtime by analyzing sensor signals and operational patterns.

Predictive Maintenance System AI tools that forecast equipment failures using sensor streams, operational logs, and historical patterns.

Predictive Pricing (Dynamic Pricing) Real-time price optimization based on shifting factors such as demand, inventory, user behavior, and competitor actions. Becomes more effective—and more defensible—with richer proprietary data.

Privacy-Preserving Techniques Methods such as anonymization, differential privacy, and federated learning designed to enable model development while protecting sensitive information.

Probabilistic Reasoning (Enterprise AI) Decision-making based on likelihoods rather than certainty—central to adopting AI responsibly.

Productivity J-Curve Erik Brynjolfsson's model of the pattern by which measured productivity appears to dip in the early years of adopting a general-purpose technology and rises only later, once the intangible complementary investments—process redesign, retraining, new business models—it requires have been made.

Prompt Injection A security attack in which malicious instructions are embedded in user input to manipulate the behavior of a language model or RAG system.

Proof-of-Concept (PoC) Limitation The common pattern in early AI adoption where pilots demonstrate technical feasibility but fail to scale due to inadequate data, integration, or cross-functional alignment.

Proprietary AI Model A model uniquely tailored to a company's workflows, optimized using exclusive data. Delivers differentiation beyond what off-the-shelf models can provide.

Proprietary Data Exclusive datasets that are difficult for competitors to obtain, often generated through scale, unique interactions, or vertical integration. Forms the backbone of most AI moats.

Q

Quantization A model-optimization technique that reduces numerical precision (e.g., FP32 → INT8) to lower inference cost and improve speed while minimally affecting accuracy.

R

RAG (Retrieval-Augmented Generation) An architecture in which a generative model retrieves relevant documents or embeddings during inference to ground its responses in accurate, contextual information.

Readiness Assessment A structured evaluation of an organization's current capabilities—including data, technology, governance, talent, and culture—to determine its capacity to scale AI. Often conducted using AI maturity models.

Real Options Analysis (AI Investments) A valuation framework that quantifies the strategic flexibility embedded in AI projects—such as the option to scale, defer, expand, or abandon—beyond traditional DCF methods.

Real-Time Monitoring Dashboards Interfaces that surface predictions, anomalies, and alerts to operations teams, especially in manufacturing and logistics.

Recommendation Engine An AI system that analyzes user behavior and preferences to suggest products, services, or content likely to interest the user, driving revenue and personalization.

Reinforcement Learning A machine learning paradigm in which an agent learns optimal actions by interacting with an environment and receiving rewards or penalties. Used in robotics, routing, trading, and other sequential decision tasks. Also: AI that learns optimal actions by maximizing rewards over time, used in route optimization, warehousing, and energy management.

Representation Bias Bias arising when training data inadequately captures the diversity of the population the model is intended to serve. Often due to skewed sampling. Also: Bias arising when certain groups are underrepresented in training data, leading to unequal model performance.

Responsible AI A holistic framework that embeds fairness, transparency, safety, privacy, and accountability into every stage of the AI lifecycle—from ideation to retirement. Also: The discipline of building AI systems that are transparent, fair, accountable, safe, and aligned with both organizational values and societal norms.

Review Board (AI Ethics Board) A governance mechanism that evaluates AI proposals for fairness, risk exposure, regulatory compliance, and stakeholder impact before approval.

Risk-Based Classification (Regulatory) The categorization of AI systems by potential harm, used to determine the level of regulatory scrutiny, documentation, and controls required.

Risk Management (AI Context) The identification, assessment, and mitigation of AI-specific risks—including bias, drift, model failure, adversarial attacks, privacy breaches, and regulatory exposure.

Risk Mitigation (AI Value Lever) Using AI to detect anomalies, forecast threats, and support high-stakes decision-making—for example, fraud detection or cybersecurity—thereby reducing exposure to financial or operational risks.

Robotic Process Automation (RPA) Software that automates repetitive, rules-based tasks traditionally performed by humans, often serving as a foundation for more advanced AI-driven automation.

S

Safety Alignment The process of ensuring that LLM outputs adhere to organizational values, legal constraints, and ethical expectations. Includes RLHF, guardrails, filters, and prompt-level constraints.

Scenario Planning (AI Economics) A strategic method for evaluating how changes in regulation, data availability, model performance, or competition could alter the financial viability of AI initiatives.

Scenario Planning for AGI Strategic preparation for multiple futures—incremental, hybrid, and transformative—driven by potential advances in general intelligence.

Scenario Stress-Testing (AI Governance) Simulating adverse or unexpected conditions—data drift, market shocks, adversarial attacks—to ensure models remain safe and reliable under stress.

Schema Validation Automated checks that ensure incoming data conforms to expected structure and types, preventing downstream failures.

Segmentation (Image Segmentation) A CV technique that divides an image into pixel-level regions, enabling precise classification of objects and boundaries for tasks like medical imaging or defect detection.

Self-Attention Mechanism A foundational component of transformer NLP models that enables them to capture relationships between words across long sequences, dramatically improving language understanding.

Sensor Drift Changes in measurement accuracy over time, requiring recalibration and model adjustments in industrial environments.

Sentiment Analysis An NLP method that detects emotional tone—positive, negative, or neutral—within text such as reviews, social media posts, or customer feedback.

Sentiment Analysis (Sales & CX) AI techniques that evaluate customer emotions and intent in text or speech to improve service quality, detect risks, and personalize engagement.

Sequencing (AI Strategy) The deliberate ordering of AI investments and initiatives to maximize momentum and value realization. Balances quick wins with longer-term transformational initiatives.

Simulation-Based Synthetic Data Artificial data generated from rule-based or physics-based simulations to supplement or replace real-world training datasets.

Skill Pathways (AI Workforce Development) Structured learning journeys—micro-credentials, internal academies, mentorship programs—that provide employees with AI-relevant skills and accelerate career mobility.

Strategic Framing The process of linking organizational goals to AI capabilities. Clarifies which problems are worth solving, what outcomes matter most, and how AI should be deployed to support enterprise priorities.

Strategic Positioning (AI Era) The deliberate selection of where and how a firm builds, deploys, and defends AI capabilities—taking into account timing, ecosystems, industry dynamics, and data assets.

Strategic Resource Allocation The targeted deployment of funding, talent, and infrastructure toward the most strategically important AI initiatives. Prevents organizational dilution and ensures sustained impact.

Streaming Architecture A processing paradigm that handles continuous, real-time data flows, enabling immediate updates to models, analytics, and decision systems.

Structured vs. Unstructured Data Pipelines Contrasting pipelines for traditional ML and LLMs: structured numeric/transactional data vs. text, code, audio, or multimodal inputs.

Supervised Learning A form of machine learning that uses labeled historical data to predict known outcomes. Widely used for forecasting, scoring, classification, and recommendation.

Sustainability (AI Infrastructure) The practice of designing AI systems that minimize energy consumption and environmental impact through hardware efficiency, model compression, renewable energy use, and optimized workload scheduling.

Switching Costs (AI Context) The economic, operational, or workflow disruptions incurred when transitioning away from an established AI system. High switching costs create durable competitive defenses.

Symbolic AI (Rule-Based AI) An early approach to AI that used hand-coded rules and logic structures to simulate human reasoning. Symbolic AI struggled with ambiguity and real-world complexity.

Synthetic Data Artificially generated data—produced by rules, simulations, or generative models—used to augment limited, sensitive, or unbalanced real datasets.

Systemic Bias See: Historical Bias.

T

Talent Analytics Predictive models that assess employee engagement, attrition risk, performance drivers, and workforce planning needs.

Talent and Capability Planning The identification and development of the skills required for AI success, including data engineering, modeling, MLOps/LLMOps, governance, and interdisciplinary business-technical translation. Often involves hiring, training, and restructuring.

Technology Roadmap (Contrasted with AI Strategy) A plan that outlines technical infrastructure or engineering activities—but does not inherently define business outcomes or value creation. A subset of, not a substitute for, an AI strategy.

Tensor Processing Unit (TPU) A specialized, high-efficiency accelerator optimized for deep learning tensor operations, enabling faster and more energy-efficient training for certain model classes.

Timing Advantage The strategic benefit of entering a market at the right moment—early enough to accumulate data and shape standards, or late enough to apply superior technology with fewer constraints.

Token-Based Pricing A granular cost model where organizations pay based on the number of tokens processed by language models. Enables precise attribution of AI costs to specific workflows.

Total Cost of Ownership (TCO) for AI A comprehensive view of long-term AI costs—including compute usage, retraining, monitoring, compliance, data acquisition, and governance—beyond initial deployment.

Transformer Architecture A neural network design that uses self-attention to process language (and increasingly multimodal) data. It powers modern NLP and large language models.

Transformative Initiative A high-complexity, high-impact AI project that has the potential to reshape business models, customer experiences, or competitive positioning. Requires significant investment and maturity.

Transparency The degree to which stakeholders can understand how AI systems operate, what data they use, and how decisions are generated. Essential for trust and regulatory compliance.

U

Unsupervised Learning A machine learning paradigm that identifies patterns, clusters, or anomalies without labeled outputs. Useful for segmentation, anomaly detection, and exploratory analytics.

V

Value-Aligned Deployment Deploying models only when they demonstrably support commercial goals, ethical expectations, and operational KPIs.

Value Creation (AI Context) The conversion of AI capabilities into tangible business results. Includes revenue growth, cost reduction, process optimization, risk mitigation, and enhanced customer value.

Value Levers (AI) The three primary ways AI generates business benefit: cost reduction, revenue generation, and risk mitigation. These levers guide strategic planning and investment decisions.

Variety (Data Diversity) The different types of data (text, images, numbers, sensor data) that enrich an AI system's understanding and improve learning.

Vector Database A specialized database for storing and searching vector embeddings produced by AI models, supporting semantic search, retrieval-augmented generation, and multimodal applications. Also: A specialized system that stores and searches embeddings efficiently, enabling semantic search, personalization, anomaly detection, and RAG pipelines.

Velocity (Data Speed) The rate at which data is generated and processed—important for real-time applications like fraud detection or dynamic pricing.

Veracity (Data Quality) A measure of data accuracy, consistency, and reliability. High veracity ensures AI models produce trustworthy outputs.

Versioning (Model / Data / Prompt) The practice of tracking all versions of models, datasets, and prompts to ensure reproducibility, traceability, and compliance in ML and LLMOps workflows.

Vertical Integration (AI-Enabled) Ownership of both data-generating assets and the operational processes that use AI. Enables end-to-end optimization, rapid iteration, and strong data moats.

Volume (Data Size) The amount of data available for training AI models. Large datasets improve generalization and accuracy.

W

Winner-Take-Most Dynamics A competitive pattern where a small number of firms capture disproportionate market share due to network effects, data flywheels, or platform dominance.

Workforce Augmentation The shift toward human–AI collaboration in which AI agents handle routine or analytical tasks while employees focus on judgment, creativity, and strategic oversight.

Workforce Orchestration (AI Agent Era) Coordinating human employees with AI agents, ensuring supervision, performance management, and escalation pathways.

Workforce Transformation (AI Era) Enterprise-wide changes to roles, responsibilities, skills, and career paths driven by the rise of autonomous AI systems and human–machine collaboration.

Workflow Augmentation The enhancement of human decision-making through AI-driven insights, predictive recommendations, and automated support functions.

Workload Optimization Strategies such as caching, distributed execution, autoscaling, and hardware selection that ensure AI systems run efficiently and cost-effectively in production.

SELECTED BIBLIOGRAPHY

Abhishek, A., Erickson, L., & Bandopadhyay, T. (2025). Data and AI governance: Promoting equity, ethics, and fairness in large language models. *arXiv.*

Agbese, M., Alanen, H.-K., Antikainen, J., Halme, E., Isomäki, H., Jantunen, M., … Vakkuri, V. (2021). Governance of ethical and trustworthy AI systems: Research gaps in the ECCOLA method. *arXiv.*

Agrawal, A., Gans, J., & Goldfarb, A. (2018). *Prediction machines: The simple economics of artificial intelligence.* Harvard Business Review Press.

"AI fairness in data management and analytics: Challenges, methodologies and applications." (2023). *Applied Sciences*, 13(18), 10258.

"AI governance frameworks and multi-level governance analysis." (2024). *AI and Ethics*, 5, 3265–3279.

Amrit, C., & Narayanappa, A. K. (2024). An analysis of the challenges in the adoption of MLOps. *Journal of Innovation & Knowledge*, 9, 100637.

Armbrust, M., Ghodsi, A., Xin, R., & Zaharia, M. (2021). Lakehouse: A new generation of open platforms that unify data warehousing and advanced analytics. *Proceedings of the 11th Conference on Innovative Data Systems Research (CIDR).*

Bade, S. (2026). From experimentation to enterprise reality: Why MLOps is the backbone of production AI. *Journal of International Crisis and Risk Communication Research*, 9(1), 17–26.

Batool, A., Zowghi, D., & Bano, M. (2023). Responsible AI governance: A systematic literature review. *arXiv.*

Bean, R. (2025). *2025 AI and data leadership executive benchmark survey.* NewVantage Partners.

Brynjolfsson, E., Rock, D., & Syverson, C. (2021). The productivity J-curve: How intangibles complement general purpose technologies. *American Economic Journal: Macroeconomics, 13*(1), 333–372.

Davenport, T. H., & Ronanki, R. (2018). Artificial intelligence for the real world. *Harvard Business Review, 96*(1), 108–116.

Dehghani, Z. (2022). *Data mesh: Delivering data-driven value at scale.* O'Reilly Media.

Dhamodharan, B. (2023). Driving business value with AI: A framework for MLOps-driven enterprise adoption. *International Journal of Sustainable Development in Computing Science*, 5(4), 1–19.

"An empirical analysis of enterprise AI adoption barriers and business value alignment." (2024). *Journal of Information Systems Engineering and Management*, 10(59s).

European Parliament and Council of the European Union. (2024). Regulation (EU) 2024/1689 laying down harmonised rules on artificial intelligence (Artificial Intelligence Act). *Official Journal of the European Union*, L 2024/1689.

Fenn, J., & Raskino, M. (2008). *Mastering the hype cycle: How to choose the right innovation at the right time.* Harvard Business Review Press.

Fountaine, T., McCarthy, B., & Saleh, T. (2019). Building the AI-powered organization. *Harvard Business Review, 97*(4), 62–73.

Giordani, J., & Zeko, R. (2024). An empirical study on enterprise-wide governance practices for artificial intelligence and machine learning. *European Journal of Applied Science, Engineering and Technology*, 2(6), 168–177.

Iansiti, M., & Lakhani, K. R. (2020). *Competing in the age of AI: Strategy and leadership when algorithms and networks run the world.* Harvard Business Review Press.

Jobin, A., Ienca, M., & Vayena, E. (2019). The global landscape of AI ethics guidelines. *Nature Machine Intelligence, 1*(9), 389–399.

Jouppi, N. P., Young, C., Patil, N., Patterson, D., et al. (2017). In-datacenter performance analysis of a tensor processing unit. *Proceedings of the 44th Annual International Symposium on Computer Architecture (ISCA)*, 1–12.

Lu, Q., Zhu, L., Xu, X., Whittle, J., Zowghi, D., & Jacquet, A. (2022). Responsible AI pattern catalogue: A collection of best practices for AI governance and engineering. *arXiv.*

McKinsey & Company. (2025). *The state of AI: Global survey 2025.*

"MLOps: The methodology and business implications of scalable machine learning operations." (2024). *International Journal of Business Intelligence Research*, 15(1), Article 358916.

Mopur, V. R. (2022). Artificial intelligence in enterprise decision-making: Bridging the gap between predictive insights and organizational strategy. *African Journal of Artificial Intelligence and Sustainable Development*, 2(2), 1–14.

National Institute of Standards and Technology. (2023). *Artificial intelligence risk management framework (AI RMF 1.0)* (NIST AI 100-1). U.S. Department of Commerce.

Pasch, S. (2025). Operationalizing AI: Empirical evidence on MLOps practices, user satisfaction, and organizational context (Preprint). *arXiv.*

Shakil, M. M., Halimuzzaman, M., Khan, A. U. Z., & Fardous, J. (2025). The impact of artificial intelligence on business strategy: Redefining competitive advantage in the digital era. *Journal of Information Technology, Cybersecurity, and Artificial Intelligence*, 2(3), 1–16.

Soni, N., Sharma, E. K., Singh, N., & Kapoor, A. (2019). Impact of artificial intelligence on businesses: From research, innovation, market deployment to future shifts in business models. *arXiv.*

Teodorescu, M. H. M., & Makridis, C. (2024). Fairness in machine learning: Regulation or standards? *Brookings Institution.*

Teoh, G., & Esther, D. (2024). AI governance in MLOps: Compliance, fairness, and transparency [Conference paper]. *ResearchGate.*

Toreini, E., Aitken, M., Coopamootoo, P. L., Elliott, K., Gonzalez Zelaya, V., Missier, P., … van Moorsel, A. (2020). Technologies for trustworthy machine learning: A survey in a socio-technical context. *arXiv.*

Wade, M., Lagodny, A., Andersen, A.-C., Avelines, C., & Plueckebaum, A. (2024). Do you really need a chief AI officer? *MIT Sloan Management Review.*

ABOUT THE AUTHOR

Dr. Samuel S. Conn brings nearly 40 years of leadership spanning the military, higher education, business, and technology sectors, combining teaching and research roles at major Carnegie-classified R1 institutions with a distinguished executive career in enterprise IT and digital and AI transformation. His scholarship and instruction span artificial intelligence, enterprise systems, big data, and cloud computing, giving him a rare blend of academic depth and executive experience in studying AI's impact on organizations.

Dr. Conn previously served as President and CEO of a national research and education network, where he led digital transformation initiatives in broadband infrastructure, AI, and cybersecurity for higher education, government, and healthcare institutions across the Northeast. His executive career also includes serving as Executive Vice President and CIO at SUNY Empire State University and Vice President for Information Technology and CIO at Southern Polytechnic State University.

A scholar-practitioner, Dr. Conn has authored more than 40 peer-reviewed publications and three books on systems design, digital learning, and knowledge management. His work has earned several distinguished and meritorious paper awards, and he was named a Technology Thought Leader by ROI-NJ in 2021 and 2022. He holds a Ph.D. in Information Systems from Nova Southeastern University's College of Computing, AI & Cybersecurity.